MAKING MOONSHOTS

MAKING MOONSHOTS

RAHUL RANA

For my father,
who imparted onto me a deep
curiosity for science + tech.

And my mother,
who gave me my entrepreneurial side.

And my brother and sister,
who always hype me up.

CONTENTS

GLOSSARY

MOONSHOT
The act of launching a rocket to the moon; an ambitious, expensive, seemingly impossible goal that will probably be unattainable.

DEEPTECH / HARDTECH / FRONTIER TECH / EMERGING TECH
Any cutting-edge combination of science and technology; an R&D-heavy project.

MOONSHOT COMPANY
A deeptech startup that makes a radical solution to address a big problem in the world; one that emulates NASA's approach to the Apollo missions.

In every interview, I asked **"what is a moonshot to you?"** These two definitions, respectively from an investor and a founder, stood out.

"I think it starts with the perception of impossibility. The idea of moonshot comes from going to the moon and the concatenated series of events that

had to occur. There are many more ways to fail and succeed. And if you think about that as a cornucopia, you know moonshot is like finding the one or two paths out of like a million that will get you there. Number two is that has to be big. There's lots of improbable things that the magnitude of their importance is just not big. It has to be meaningful. Breaking the Earth's atmosphere and landing on an orbiting satellite was historic. And so, I think it has to be improbable and it has to have a magnitude of importance. That's huge. So, basically low frequency, high magnitude."

"There's no such thing as high-reward, low-risk in the world. It's a company predicated on the idea that there is a 90 percent chance that it fails. If it succeeds, it's one of the most important companies on Earth. But I think if there's less than 90 percent chance of failure, the company probably won't be that valuable."

1

INTRODUCTION

REDEMPTION AND REVOLUTION

"We are not here to curse the darkness, but to
light the candle that can guide us through
that darkness to a safe and sane future."

—JOHN F. KENNEDY

CATCHING UP TO RUSSIA

September 12, 1962. It was a hot day at Rice Stadium under
the pressure of the boiling Cold War.

Surrounded by thousands of inspired scientists, leaders,
politicians, and students, he addressed the attendees with a
literal target on his back.

The sharp, Bostonian pastiche. The young, televi-
sion-friendly disposition with a great head of hair. A symbol

of vitality, charisma, charm; of grace, of romantic idealism, of mystical passion. A master orator, a powerful thinker, a captivating commander-in-chief.

> *"We choose to go to the moon in this decade and do the other things not because they are easy, but because they are hard, because that goal will serve to organize and measure the best of our energies and skills, because that challenge is one that we are willing to accept, one we are unwilling to postpone, and one which we intend to win, and the others, too."*[1]

President John F. Kennedy's Moon Speech was an energizing call-to-action that galvanized the country into adopting his goal of safely reaching the moon by the end of the decade. Filled to its brim with astronomy metaphors and patriotism, the talk portrayed the inevitable push for space exploration and why America should be at the forefront of it. He noted how grand challenges, such as a moon landing, bring out the best of us—courage, ingenuity, and determination. Inadvertently, Kennedy ushered in a new movement of American innovation and invention.[2] Despite history having innumerable examples of this style of scientific and technological progress, JFK popularized the term "moonshot."

---

1   John F. Kennedy, "John F. Kennedy Moon Speech (1962)," AIRBOYD, speech presented on September 12, 1962, YouTube video, 17:47.

2   Ibid.

John F. Kennedy at Rice University[3]

Yet, under the empowering, feel-good surface of the speech was a PR pitch for public support. The truth was Kennedy needed a political win, especially after the Bay of Pigs fiasco. The Soviet Union already had countless firsts, including the launch of Sputnik and Yuri Gagarin's first-ever human spaceflight.

---

3   History in HD, "JFK at Rice University, Houston, United States."

The Moon Speech came six weeks after that, with America desperate for a demonstration of military strength and scientific prowess.[4] In fact, both NASA's creation and the Apollo missions were "catch-up and overtake" efforts by America, although NASA came from the Eisenhower Administration.[5]

Nonetheless, Kennedy admitted it would not be an easy road ahead. He knew his goal was incredibly audacious, and probably impossible. In the '60s, going to the moon not only meant sustaining travel over nearly 240,000 miles, but also inventing new metal alloys and going where no human has. There were no spacesuits, space food, or computers suitable for space navigation at the time.[6] Not to mention, NASA needed a functional rocket: one that could withstand insane temperatures and conditions, reach speeds of over 25,000 miles per hour without shattering, and carry all the necessary equipment for propulsion, guidance, control, and communications.[7] But, Kennedy chose to be an optimist and reimagined the potential of scientific research, technological ingenuity, and economic resources. He knew one thing was clear. To invent and innovate upon all of those things safely and successfully, we needed one thing—"We must be bold," he said.[8]

4   Jennifer Frost, "Who Really Won the US-Soviet Space Race?" *Newsroom,* July 19, 2019.

5   "The Decision to Go to the Moon: President John F. Kennedy's May 25, 1961 Speech before a Joint Session of Congress," NASA History Office, updated October 29, 2013.

6   Astro Teller, "We Choose to Go to the Moon," *X, The Moonshot Factory* (blog), *Medium,* July 23, 2019.

7   Alexis C. Madrigal, "Moondoggle: The Forgotten Opposition to the Apollo Program," *The Atlantic,* September 12, 2012.

8   John F. Kennedy, "John F. Kennedy Moon Speech (1962)," AIRBOYD, speech presented on September 12, 1962, YouTube video, 17:47.

Naturally, people thought he was out of his mind. Some thought it was a complete waste of money, and others thought it was fundamentally unfeasible. In fact, public opinion polls conducted around the time of the Apollo missions found more than 50 percent of Americans believed the lunar missions were not worth the cost.[9] Kennedy met with backlash as he focused on the next frontier while his own frontier had a race war brewing.

Even the brightest scientists were baffled and disheartened at the impossibility of the idea. A *Science* academic journal poll of 113 non-NASA scientists found 110 out of 113 of them did not believe in the Apollo missions, as they thought America was unreasonably rushing toward a man on the moon.[10] This presented a conflict of interest: can we really go to the moon, or is Kennedy crazy?

In 1962, a moon landing was this utterly unimaginable, science-fiction-sounding concept that challenged the fabric of humanity. It was heresy. First of all, we had no capabilities. Second, why not fix unemployment? The lack of health care? Racial equality? The incessant fear of being bombed by Cuba? People were confused, but it was hard not to smile about reaching the lunar surface.

In the face of the opposition, the impassioned strategist pledged to pour all of his effort into making the impossible possible. He mobilized the government, which contributed upward of $25 billion to Apollo. He tapped NASA's Wernher von Braun to take charge as the chief architect behind the Saturn V launch vehicle. He worked with MIT's Instrumentation

9   Alexis C. Madrigal, "Moondoggle: The Forgotten Opposition to the Apollo Program," *The Atlantic,* September 12, 2012.

10   Ibid.

Lab to build the computer systems for the rockets. Finally, continued measures from the following two presidencies upheld the dream and led to the successful Apollo 11 mission.

ENTREPRENEURSHIP

Fun fact—the term "moonshot" was actually being used before the '60s. The first relatively widespread instance of it was actually in baseball. According to the Los Angeles Dodgers media coordinator Cary Osborne, announcer Vin Scully coined the term in 1959 when Dodgers All-Star Wally Moon hit home runs over the 42-foot left field wall in the L.A. Memorial Coliseum.[11] Nicknamed the "Moonibrow," Moon adapted to an awkwardly-situated baseball diamond by changing his swing to an unconventional, uppercut chipping motion. In turn, the media picked up his towering home runs and referred to them as "moon shots."[12]

Despite the transition in meaning from baseball to space exploration of the portmanteau "moonshot," both Moon and Kennedy represent what one is. The former is about creative execution and lofty goals—42 feet, to be specific—while the latter is about a zealous outpouring of resources toward an intractable problem. Wally was hitting out of left field (literally) with a swing that broke the status quo, while Kennedy channeled the exigency of the Space Race and the Cold War into a seemingly impossible technological feat. Wally's unibrowed and uncombed style, not to mention his unusual swing, aesthetically describes the chaos in "moonshotting," while Kennedy's "Cold

---

11 Cary Osborne, "One of the First Los Angeles Dodgers Stars, Wally Moon Passes Away," *Dodger Insider* (blog), *Medium*, February 10, 2018.

12 Alex Davies, "Why 'Moon Shot' Has No Place in the 21st Century," *Wired,* July 16, 2019.

Warring" was quite conventional at the time. Anything to beat Russia, after all.

Kennedy's nearly decade-long plan symbolizes something more than the giant leap for mankind that it was. It was the first successful moonshot, a resounding victory for the American industry itself. Accordingly, the Apollo 11 mission was a defining moment in the history of entrepreneurship. Sure, it was a government-funded effort from a government agency with a government workforce. But it imparted on the world this notion of doing big R&D things that also positively impact the world.

There are startup founders out there who follow a similar course of action. They employ strategies and mindsets similar to Kennedy's. They take the processes similar to those behind the Apollo program and apply it to business for the benefit of all of humanity. The result: a world of exponential progress.

These are what I call **"moonshot companies."**

Forget Milton Friedman's manifesto "the social responsibility of business is to increase its profits," usually at the expense of humans, the environment, or morality.[13] These are the startups that achieve bold goals in three parts: building ground-breaking technologies, leveraging them in radically creative solutions, and solving deep, systemic problems in the world.

They are the products of futurism and exigency and failure and uncertainty. They are risky to pursue, technically difficult, and vastly rewarding for all of society. They chase abundance, change, and 1,000x returns—not just 10 percent improvements.[14]

---

13 Milton Friedman, "A Friedman Doctrine—The Social Responsibility of Business Is to Increase Its Profits," *The New York Times archive,* September 13, 1970.

14 Astro Teller, "We Choose to Go to the Moon," *X, The Moonshot Factory* (blog), *Medium,* July 23, 2019.

They revolutionize an industry or workflow of society while also spreading massive positive externalities into the world.

The only bounds these startups have are those placed by us.

Yet, instead of rhapsodizing, let me put it in realistic terms. I would say any impactful deeptech initiative can be considered a moonshot. It's more of a three-part spectrum rather than a concrete framework. Deeptech entails any cutting-edge, R&D-heavy project at the crossroads of applied and emerging sciences and technologies.

There is no hack to building companies such as these—you have to go out there and stay persistent.

However, a secret sauce can be used to increase the probability of success, namely key qualities of the mavericks who launch them—people such as Elon Musk (Tesla), Reshma Shetty (Ginkgo Bioworks), and Thomas Reardon (CTRL Labs). One is they strive for massive change rather than incremental steps forward. Another is they celebrate failure and navigate entrepreneurial complexity. Those principles are the demarcating lines between a fraud and a real fixer. Moonshots are characterized by a checklist of surface-level fundamentals, yet they can be as philosophically deep as maximizing human potential and gaining a better understanding of the universe. They are the closest thing we have to magic.

We need more of these companies because our closeted dystopian society is packed with problems such as natural disasters, food scarcity, unsustainable energy consumption, vulnerable infrastructure, pollution, disease, environmental destruction, unemployment, unnatural death, expensive health care, aging populations, and so much more. However, in the face of these issues come entrepreneurs for everything from the increasing number of vehicle-related deaths

(Waymo) to the inability to store massive amounts of data (Catalog Technologies) to nuclear waste clean-up (Kurion).

On top of that, the beauty of moonshots is they aim to be financially sustainable. At their core, they are still businesses. Admittedly, some are nightmares—Tesla's financials are quite questionable, or Kleiner Perkins's failed Green Fund. But there is a difference between the common definition—an unsustainable, impractical goal—and what I speak of. The combination of profit and purpose is what drives these initiatives. They are the ones that make the future. After all, as famed computer scientist Alan Kay said, "the best way to predict the future is to invent it."[15]

THE FUTURE, TODAY

So many industries are building the future: artificial intelligence, machine learning, deep learning, quantum computing, genomics, human-computer interaction, autonomous transportation, augmented and virtual reality, space technology, synthetic biology, bioinformatics, agriculture technology, manufacturing, robotics, nanotechnology, neurotech, cleantech, energy, IoT, drones, materials science, and so many more combinations of the hard sciences and technology. Not to sound naive, but it's *so cool.*

A concession must be made, however. Traditional startups are not any less important, as they may as well come across a breakthrough. Similarly, what's considered deeptech may become more of a traditional company over time (think of early internet companies versus the commoditized digital landscape of now). Relatively speaking, the "regular"

15 Alan Kay, "Early Meeting in 1971 of Parc, Palo Alto Research Center, Folks and the Xerox Planners," (meeting, Xerox Palo Alto Research Center, Palo Alto, CA, 1971).

companies have a larger probability of success. They innovate along existing workflows and provide painkillers to certain problems in the world. Deeptech does that, except with immense missions within uncharted territory. Thus, on the axes of commercial and technical risk, traditional startups *might* be safer bets, while moonshots are a recipe for a tough road ahead.

But that is where their value is. A big thing stopping people from pursuing these projects is the notion that these types of endeavors are too idealistic, too imaginative. Some think it's just a marketing ploy. Others react adversely to the perceived risk and low potential for success. They fear failure.

The exciting fields exist. It's just the endeavors within them seem to be too impractical. People fall back to working on traditional startups, often undervaluing the opportunity cost of not doing the most far-fetched thing they could be doing. Moonshots do not imply losing track of all practicalities. They may look like traditional companies, but with their outsized ambitions and longer-term visions comes output many yearn for.

At the end of the day, if people continue to stray away from pursuing radically creative breakthroughs, we will never make exponential progress in society. We will be ordinary. Global issues will go unsolved. The vast majority of us will not achieve our human potential. We will hit a stagnation. If we don't build, then who will?

That is currently what plagues America and the world. We've drifted from our innovative roots toward an individualistic society of misaligned incentives and underutilized science.[16] And, it shows through our minimal rate of progress in some deeptech industries.

---

16   Richard Ngo, "Thiel on Progress and Stagnation," LessWrong, July 20, 2020.

So, why stay comfortable in our current rate of innovation and invention? Why not jumpstart solutions to overarching crises such as extreme energy demands, worsening health conditions, or exploiting nature? Why not make more organizations that are hacking the gut-brain axis (Kallyope), creating carbon-neutral fuel from seawater (Project Foghorn), and controlling machines with your mind (CTRL Labs)?

We've been doing this for ages. It's in our blood. Humanity is defined by overcoming Malthusian traps, or catastrophes that occur with skyrocketing population growth. Yet, I interpret that differently. It's rather a testament to how as we evolve, so too do the challenges we face. Despite the increasing complexity of what we deal with, we still build grand civilizations, spark dozens of golden ages, and start intellectual revolutions. We adapt. We find a way to win. I'm here to say we need more of this.

It's natural to shy away from it all, so let's explore the journey in making a world-changing company: the philosophies, the mindsets, the strategies. I hope, in reading this book, you will discover moonshot thinking and actively apply it to your own life. We will then put that mentality into tangible action and end with what to do on an ecosystem level. While I portray this in the context of frontier tech startups, the universal lessons can be used for anyone's personal goals. Ultimately, I hope to instill the same zeal in you that I have for important breakthrough enterprises by demystifying the nonobvious aspects of founding one. My work will be complete if I can inspire at least one reader to start working on their boldest idea.

One more successful moonshot has the potential to drastically reduce any aspect of human suffering and get us out of the stagnation we face. One more moonshot can push the course of technological progress and societal progress in turn. A movement starts with one. When something previously thought of as inconceivable becomes a reality, it spurs action individually and collectively levels.

This is for those who love radical entrepreneurship: founders, venture capitalists, contrarians, disruptors, futurists, inventors, scientists, researchers, problem-solvers, leaders, students, policymakers, and more.

It's time to make moonshots.

2

MINDSET

LUNAR ASPIRATIONS

—

THE BRAIN BEHIND THE MOON

Every single day, tragedies around the world bombard us with anxiety. Bad news is now the norm. We are at the point where people are desensitized to the jarring realities of life.

I'm writing this in a pandemic. Yet, around the globe, we've got human rights violations, food deserts, environmental destruction, nonrenewable resources, minimal education, crumbling infrastructure, broken healthcare systems, rampant criminal activity, and so much more. I don't need to be the one to tell you—just check the news. It sometimes seems like these crises are out of our reach for us to even make a dent in them.

The thing is bad news sells and it always has. Connectivity has just made it more visible; it lights a fire under many people. We just need to better utilize and harness this motivation. We need to inspire people to effectively solve these vast issues while also advancing scientific, technological, and human progress. It's a huge task, but I realized that at the core of it is the mindset. I'm playing the long game. I know once you have the mindset, you can do anything. You can think and act like a moonshot founder and, in turn, pursue being one on your own.

Hence, over the course of this entire book, we're going to explore the nonobvious foundation of how some brilliant operators work and then put it into action. First, we'll lay the basics and then compound them with philosophies and mental models to paint a comprehensive picture of the "moonshot mindset."

Before all of that, it should be known what works for some may not work for others; there is no perfect way of thinking. There is no hack on how to suddenly impact the world for the better, nor is there one on how to emulate the mind behind people such as Anousheh Ansari, Laura Deming, and Sam Altman. Although trust me—you *do* want to be like them. Ansari is an escapee of the Iranian Revolution turned astronaut, co-founder of Prodea Systems, and now CEO of XPRIZE, the gold standard of deeptech competitions. Deming is an all-around legend in the longevity and cell biology industry after working in four different labs and founding the Longevity Fund, a multimillion-dollar investment firm for age-related disease breakthroughs. Finally, Altman is a luminary within the startup arena, previously being the president of Y Combinator and currently CEO for OpenAI. You get the point.

One thing that is possible is to learn the best practices from founders on the frontiers of science and technology, whether it is in making decisions, being a good steward of innovation, or constructing a culture of creativity.

As we discuss the psychology behind emerging tech startups, I will purposefully skip the obvious things. We all know you need passion in your work and you have to be determined, resourceful, and whatever other plain-yet-necessary traits. What separates moonshot founders from others is this collection of lesser-known qualities and thought processes that guide them in the entrepreneurial complexity at the crossroads of deeptech and societal impact. This entails so much, from radical creativity to diversity of thought. But I've boiled it down into the most prominent aspects.

The first step: let's aim for the moon.

THE BLOOD OF A MOONSHOT

———

"Here's to the crazy ones. The misfits. The rebels. The troublemakers. The round pegs in the square holes. The ones who see things differently. They're not fond of rules. And they have no respect for the status quo. You can quote them, disagree with them, glorify or vilify them. About the only thing you can't do is ignore them. Because they change things. They push the human race forward. And while some may see them as the crazy ones, we see genius. Because the people who are crazy enough to think they can change the world, are the ones who do."

—STEVE JOBS

POWERING PROGRESS

Our most marvelous invention was the invention of invention itself.

Our most archaic predecessors, such as *Homo erectus* and earlier humans, were largely stagnated in terms of evolution and progression. Their lifestyles stayed static for more than 1.5 million years as they made slight divergences from the lineage of apes.[17]

It took many years to evolve into *Homo sapiens*, but the critical factor is there was not just a monumental shift in cognition and culture. The previous linear advancement suddenly became one of nonlinearity, and it is theorized this is because of the advent of communication and the spread of ideas. With the resulting collaboration, instruction, and experimentation came the emergence of creativity and invention as we know it.[18]

Fast forward to now, and we see many parallels to our earliest ancestors.

We now live in this hyperconnected society with instant communication being the norm. Everyone can process multiple ideas and make decisions. Back in the 1900s, we peaked in human ingenuity in terms of breakthroughs—innovations and inventions. That meant discovering transistors, the internet, space technology, chemical elements, and so much more. With all this compounding, we *should* stand on the shoulders of giants who figured out things so we don't have to.

Yet in recent times, those terms have taken a turn for the worse. Now, innovation means maximizing advertising revenue, new flavors for chips and soda, and even making spreadsheets. That's right—the once $27 billion German bank Wirecard AG stressed how innovative their company

17   Keith Frankish, "Our Greatest Invention Was the Invention of Invention Itself," Psyche, June 24, 2020.

18   Ibid.

was, but in a recent exposé of their criminal underpinnings, their version of "machine learning" was found to be just accountants and Microsoft Excel.[19] That's only one humorous example, however.

The glass-half-empty people argue we've stagnated in our rate of scientific and technological progress, and the glass-half-full people will say we're doing incredible things. I fall into a mix of those categories. While we are achieving some breakthroughs, the rate at which we are doing so is not keeping up with the barrage of problems that strikes us as time goes on—almost like the Malthusian catastrophe, in more than just food supply. Like the *Homo erectus* versions of us, we focus mostly on linear progression within our contemporary times. We aren't at the rate we should be at, and we are definitely not doing enough to match the sheer outburst of complex, global problems.

There's so much working against the quest to solve these crises—politics, the lack of R&D funding, outdated education, individualistic culture, unbalanced incentives, and more. But, the biggest thing most of us lack is the mindset behind making progress. We lost that sense of ingenuity, of advancement, of betterment for all. Altogether, I call this "radical creativity."

There are some pockets of this genius way of thinking. What was once Bell Labs and Xerox PARC is Google's X Moonshot Factory and Telefonica's Alpha. You have government labs, startups, and universities out there engaging in cutting-edge domains. While they're doing ambitious research, some are also scaling it and bringing it to the masses. Thus, instead of the polarization of progress, all of humanity

---

19   ColdFusion, "The Wirecard Fraud - How One Man Fooled All of Germany," ColdFusion, August 6, 2020, YouTube video, 17:05.

can feel the benefits. Like these organizations, we must all invest into futuristic domains. To foster more moonshots, we must reintroduce *everyone* to this lesser-known way of thinking from idea to product.

My thesis is clear: radical creativity is what pushes forth deeptech.

CREATIVITY SQUARED

First, we must establish what normal creativity is. It is tough to define, but it typically is described as thinking in an original, artistic, and combinatorial fashion. I define it as a mental framework and sublevel of consciousness unlocked by varying levels of passion, constraints, and cognition that drives oneself to create, connect, and continuously update ideas while acting upon them. At its best, it is a contagious, extraordinary mental high that inspires idea generation, risk-taking, artistic expression, knowledge sharing, emotional intelligence, and more.

Regular creativity is about making connections and associations with a linear assumption of the future. It is relatively more subjective, such as in art, problem-solving, or figuring a way out of constraints. It means making new things out of existing things.

Radical creativity is on another level, although it still requires the first one. This is the mindset that spurs groundbreaking inventions. It assumes a nonlinear future or one with exponential possibilities instead of one that evolves incrementally. It's radical because you make new things out of new things in a forward-thinking context.

Radical creativity is the blood of a moonshot.

Mihaly Csikszentmihalyi believed in a similar concept. The famed psychologist behind the concept of a "flow state" applied his work to what differentiates inventive people versus those otherwise. In studying nearly 100 people legendary in their respective fields—including fourteen Nobel Prize winners, musicians, mathematicians, economists, paleontologists, and so many more—he narrowed the findings down to ten traits.[20] It's safe to say they are moonshot people. In *Creativity: Flow and the Psychology of Discovery and Invention,* Csikszentmihalyi lists:

1. *Creative individuals have a great deal of physical energy, but they are also often quiet and at rest.*

2. *Creative individuals tend to be smart, yet also naive at the same time.*

3. *A third paradoxical trait refers to the related combination of playfulness and discipline, or responsibility and irresponsibility.*

4. *Creative individuals alternate between imagination and fantasy at one end, and a rooted sense of reality at the other.*

5. *Creative people seem to harbor opposite tendencies on the continuum between extroversion and introversion.*

6. *Creative individuals are also remarkably humble and proud at the same time.*

7. *Creative individuals to a certain extent escape this rigid gender role stereotyping [of "masculine" and "feminine"].*

---

20 Mihaly Csikszentmihalyi, Creativity: Flow and the Psychology of Discovery and Invention (New York: Harper Perennial, 1997), chap. 3, Apple Books.

8. *Creative people are both traditional and conservative and at the same time rebellious and iconoclastic.*

9. *Creative persons are very passionate about their work, yet they can be extremely objective about it as well.*

10. *The openness and sensitivity of creative individuals often exposes them to suffering and pain yet also a great deal of enjoyment.*[21]

I believe all of these traits describe the minds behind the most impactful deeptech founders out there. Ironically, I made remarkably similar findings after the countless primary and secondary interviews I've completed, all of which happened before I found the study from Csikszentmihalyi.

Nonetheless, the first thing I realized is very similar to the structure of Csikszentmihalyi's traits—notice how they are all opposing qualities. Thus, a pattern that emerged was moonshot operators balance inner contradictions, which F. Scott Fitzgerald encapsulated: "The test of a first-rate intelligence is the ability to hold two opposed ideas in mind at the same time and still retain the ability to function."[22] I believe radically creative thinkers not only are those who can consider two opposing ideas without internal deterioration, but are those who thrive and find harmony in the clash between contradicting things. Deeptech investor, Steve Jurvetson, once said founders must be extremely self-confident yet

---

21 Ibid.

22 F. Scott Fitzgerald, "The Crack-up: A Desolately Frank Document from One for Whom the Salt of Life Has Lost Its Savor," *Esquire Classic,* February 1, 1936.

humble (like number 6 in the above list).[23] In a Y Combinator interview, Elon Musk said founders must be extremely pessimistic yet optimistic (like number 4).[24] Astro Teller of the X Moonshot Factory wrote radical creativity comes from being passionately dispassionate (like number 9).[25]

Furthermore, I found moonshot founders derive their unique inventiveness from other pathways.

First, they have a vibrant passion for the issue at hand, but systematically take it one chunk at a time with precise execution and a clear mind (like number 1). This leads to channeling charisma into the work but also being an introspective, reflective leader (like number 5). Also, it leads to hyper-fluency. I heard this term from 1517 Fund co-founder Danielle Strachman, who describes it as having an insatiable curiosity to relentlessly learn more about the given field.

Next, they are incredibly technical and pragmatic while also being weird, unconventional, and childlike (like number 2). A byproduct of this is asking "Why?" "How?" and "What?" to everything. This is essentially about having an unwavering skepticism about generally assumed truths and knowledge—whether that means being rebellious or ignoring the popular opinion (like number 8). On the other hand, it also entails not waiting to be told what to do. Radical creativity spurs from taking initiative and doing disruptive things you're not supposed to do (like number 7). Similarly, it also means being "responsibly irresponsible: or profoundly

23 Elevate SIX, "Steve Jurvetson: Investing in Moonshot Companies," Elevate SIX, January 9, 2020, YouTube video, 33:04.

24 Y Combinator, "Elon Musk: How to Build the Future," Y Combinator, September 15, 2016. YouTube video, 19:32.

25 Astro Teller, "Tips for Unleashing Radical Creativity," X, The Moonshot Factory (blog), Medium, February 12, 2020.

understanding a concept, but also moving fast and being resilient to change (like number 3).[26]

It also arises from embracing empathy and sympathy for suffering victims or markets with unmet needs. These operators genuinely want to see humanity flourish (like number 10)! With this wide-angle view comes the notion of being mindful of complexity—or understanding serendipity, the consequence of their actions, and the entropy of ideas, especially in the global sphere.

Finally, such entrepreneurs adopt a growth mindset, or the everlasting desire to improve oneself morally, physically, and intellectually. This may come from an internal motivation or from growing up with challenges such as disabilities, poverty, or anything that constructs a chip on their shoulder.

This is just the foundation. The key thing to note is while these are the best practices from the fifty-plus accomplished experts I've interviewed and researched, they are by no means comprehensive. The truth is creativity, radical or otherwise, works in different ways for different people. It responds uniquely to interactions, environments, cultures, and timing—just like what happened for early humans.

I'm not saying, "Just be radically more creative!" Anyone can do it: ask uncomfortable questions. Look to your heart. Experience more. Inspire yourself. Question assumptions. Do the not-so-sexy things. Brainstorm freely. Allow yourself to think bigger. Align yourself with a lofty goal. Operate at the extremes. Find a means of expression. Develop a conducive ecosystem. Aim for exceptional performance. And, recursively reflect.

---

26 Gimbal_v2.0_X (PDF), X Moonshot Factory, accessed August 29, 2020.

Many of the people who create world-changing impacts have these common factors. Progressing humankind as we know it is no easy task. But, if people embrace these traits to any extent, I wholeheartedly believe it will develop more moonshot founders by psychologically maximizing everyone's potential.

If creativity is the lifeline of a moonshot, then our survival as a species depends on it.

Let's unleash that energy.

ICONOCLAST
INVENTORS

———

"It's actually pretty easy to be contrarian.
It's hard to be contrarian and right."

—REID HOFFMAN

SCIENCE AND STARTUPS

Imagine devoting your life's work to unearthing a fairly complex concept within neuroscience, and even getting a principle named after you, but having that work almost invalidated by a newcomer to the space.

That sucks.

This is exactly what happened to neurophysiologist Elwood Henneman. The Henneman Size Principle states in motor recruitment, the neurons that control muscles act in groups according to the magnitude of the force needed for an action.[27]

---

27 Lorne M. Mendell, "The Size Principle: A Rule Describing the Recruitment of Motoneurons," *Journal of Neurophysiology* 93 (Summer 2005).

Essentially, it was believed you cannot turn individual neurons on or off and you can only fire off a group of them with the number determined by a movement's strength. Moonshot founder, Thomas Reardon, and his team at CTRL Labs disproved that principle by building a noninvasive hardware solution that connects one's nervous system to a computer or machine—effectively allowing a user to control the machine by having an intent to act. Upon isolating each fired neuron, the signal that would go to a muscle can be translated and expressed on a brain-computer interface. Yes, CTRL Labs made mind-control a reality.[28]

Think of the Force in Star Wars. By merely thinking about a certain action, those who can wield the Force can control an external entity—whether it's an object or an entire person. Reardon claims you can do just that and even learn to drive an F1 racecar in under ten minutes. Reardon figured out a way to "escape the boundaries of your body" by fundamentally granting a user more control over their life.[29] His motivation?

"I'm working on human happiness. I'm working on empowering people. I'm working on giving people new, extraordinary abilities... I didn't want to create technology that was for 100,000 people or a million people. I wanted to create technology that was for 8 billion people," says Reardon.[30] While he wants to prove the "impossible" wrong and build a high-leverage project, he is all about giving more

28 Josh Wolfe, "Episode 1: The Magic of Your Mind; CTRL-labs' Neural Interface Unlocks Human Potential," Lux Capital, August 1, 2019, YouTube Video, 6:28.

29 Ibid.

30 Demetri Kofinas, "Neural Interfaces and the Future of Human-Computer Interaction | Thomas Reardon," interview of Thomas Reardon by Josh Wolfe. *Hidden Forces,* March 11, 2019, audio, 1:04:57.

autonomy to people's lives. That's quite fitting considering that he built Internet Explorer in the '90s. He knows how to impact and empower a billion people—in typical moonshot fashion. It's also worth noting that in only four years since its founding in 2015, Facebook acquired CTRL Labs for more than $500 million.[31] That's just one example of how such a company can cause a paradigm shift in scientific fact, actualize the potential to benefit billions, and monetize the breakthrough.

CTRL Labs' technology can be the operating system of how we fundamentally interact with all technology in the future. But, that's still to come. For a more established example, it's worth mentioning the story of Genentech— the first modern biotechnology company, founded in 1976 by Herbert W. Boyer and Robert A. Swanson. After sitting on nearly thirty years of pure bench science around recombinant DNA, protein creation, and more, Genentech commercialized the molecular biology into drugs, with its first one being insulin. People thought it was impossible for the longest time. It was highly untraditional and unexpected at the time. Yet, fast forward to 1980, and Genentech became the first biotech company to go public while raising $35 million on its IPO ($114 million today). Then jump to 2009, and Genentech was acquired by Roche Holding AG in 2009 for $46.3 billion.[32]

Not only did Genentech launch some of the most significant drugs to date, considering how many lives are saved or safeguarded by them, but they also sparked the biotech

---

31 Salvador Rodriguez, "Facebook Agrees to Acquire Brain-Computing Startup CTRL-Labs," *CNBC*, September 23, 2019.

32 Jonathan Smith, "Humble Beginnings: The Origin Story of Modern Biotechnology," *Labiotech.eu*, July 2, 2019.

industry as a whole. They changed the scientific, technical, business, and legal landscapes around biology while ushering in a new era of venture capital and deeptech.[33] Just think about the scale of their impact. Their DNA sequencing technology created growth hormones for those with deficiencies, anemia treatments for chronic kidney failure, and measures to prevent infections in cancer patients undergoing chemotherapy.[34] The list is quite extensive. At the same time, they jumpstarted biotech investing by their massive returns for venture firm Kleiner Perkins, among others. They changed the game for patent laws in molecule discovery. Around the time of its IPO, President Jimmy Carter signed into law the Stevenson-Wydler Technology Innovation Act, which incentivized government-sponsored breakthroughs to be commercialized more often, and the Bayh-Dole Act—which protected research institutions' intellectual property even if it came from federal funding.[35]

Genentech paved the road for all modern-day therapeutics, drug discovery, and genomics companies, which are legitimately expanding human longevity and improving billions of people's livelihoods. It is, without a doubt, a deeptech initiative we all needed.

Clearly, moonshot companies are both visionary and profitable. But you can't have both and be a conventional strategist. They are run by contrarian thinkers—those who think in this different, radically creative way that most people don't. I break contrarianism down into three subtopics: scientific and technical paradigm shifts, thought processes, and personal qualities.

---

33   Ibid.

34   Navin Chaddha, "Biology as Technology Will Reinvent Trillion-Dollar Industries," *TechCrunch*, September 17, 2019.

35   Henry T. Greely, "The Two Months in 1980 That Shaped the Future of Biotech."

ON PARADIGM SHIFTS

Moonshots are the contrarian bets that prove to be successful. They go against the grain, proving something is possible when the masses think otherwise. Like Reardon, Boyer, and Swanson, all of whom looked where others didn't, contrarian founders reject popular opinions or assumptions and have the potential to spearhead an exponentially impactful progression of science as others tend to make tiny advances. Conviction in the face of naysayers, discoveries in light of haters—it's all in the nature of scientific revolutions.

This concept is from physicist and scientific philosopher, Thomas Kuhn, author of *The Structure of Scientific Revolutions.* In his book on the nature of scientific progress, he talks about the relationship between normal science and paradigm shifts that change underlying models of thinking within a field. Kuhn argues scientists, researchers, and practitioners do their work under certain paradigms, or assumptions, frameworks, and generally accepted truths. At this time, people build off sequential sets of ideas and add to the general pool of knowledge.[36]

Then, when unexplainable phenomena and one too many anomalies accumulate, the practitioners find themselves in unfamiliar grounds, which then leads to perspective changes. Kuhn found "the rules of normal science become increasingly blurred" as the herd attacks contrarians and those who question anomalies.[37] New ideas challenge old ones. The new standards are met with scrutiny. A conflict arises.

---

36 Thomas S. Kuhn, *The Structure of Scientific Revolutions: 50th Anniversary Edition* (Chicago: University of Chicago Press, 2012), Apple Books.

37 Thomas Kuhn: The Structure of Scientific Revolutions," *Farnam Street* (blog), *fs.blog,* accessed August 16, 2020.

In fact, Albert Einstein said it best when he once described Planck's quantum theory: a scientific revolution is "as if the ground had been pulled out from under one, with no firm foundation to be seen anywhere, upon which one could have built."[38] Yet, the thing about the unexpected, sparring ideas is the new understandings are typically inevitable. And, they're driven by contrarian thinkers whose breakthroughs cause a massive shift in thinking for the rest of the field. It's not just the destruction of the previous way of thinking, but a new way of perceiving the world and the given field.

According to Kuhn, "paradigms provide scientists not only with a map but also with some of the direction essential for map-making," and it grants them new methods, theories, and imaginations.[39] This presents another notion: the pool of scientific knowledge is deepened more by these disruptive breakthroughs as opposed to the relatively smaller innovations that occur in between each advancement. I'm not saying these incremental improvements are futile—that would be foolish and ultimately incorrect. What I am saying is progress is made when contrarian viewpoints check and balance the preconceived knowledge of a given domain.

I mention all of this because Kuhn presents a readily similar structure to that of deeptech development in entrepreneurial terms. Breakthroughs for an exponential societal benefit, all driven by unconventional thinkers—this is exactly what a moonshot is. They are the anomalies. They start revolutions. It all starts with the contrarian founders.

---

38 John S. Ridgen, "Einstein's Revolutionary Paper," *Physicsworld*, April 1, 2005.

39 "Thomas S. Kuhn, *The Structure of Scientific Revolutions: 50th Anniversary Edition* (Chicago: University of Chicago Press, 2012), Apple Books.

INDEPENDENT THINKING

Similarly, with rejecting herd mentality or groupthink for the greater good comes independent thinking. I've noticed moonshot founders sometimes engage in contrarianism in terms of thinking on their own and executing upon it.

When you're operating on the bounds of human capability and the frontiers of deeptech, you are inherently going at something very few people have before. Many have failed. It's the "frontier" for a reason. This requires one to really make personal observations and decisions without the desire to appease anyone else or get along with them.

Many of these scientific entrepreneurs push the boundaries of a generally accepted reality. Space launches were considered a waste after the Apollo missions, and they were even shunned, yet Beal Aerospace flipped the script by making the largest liquid rocket engine back in 2000.[40] People resolved to have animal meat and not be forced into vegetarianism, so Uma Valeti and Memphis Meats came along with cell-grown meat. There's a reason why people believe moonshots are impossible—it's challenging stuff. But that can be tackled by clear-headed, independent thinking because it entails questioning accepted deeptech assumptions. Thus, being a rebel scientist or founder presents an advantage: you see what other people don't and have the ultimate opportunity to take the proprietary insight and make a company out of it.

The value of being contrarian comes from the fact that these companies navigate uncertainty better than any established organization in the same space. Similar to what Kuhn said, moonshots are the anomalies that blur the lines

---

40 "Beal Aerospace Fires Largest Liquid Rocket Engine in 30 Years," *Spaceref,* March 4, 2000.

of incremental, normal science and prepare the industry for radical change. Therefore, moonshots thrive within that uncertainty, or in Kuhn's words, the crisis before the paradigm shift. They arise from it and lead the field toward this new way of thinking and operating, but they do so with an inextinguishable drive that propels them forward in the face of intellectual and sometimes physical opposition.

Deeptech investor, Josh Wolfe, says "contrarianism isn't about whether the crowd is wrong, but about exploring in what highly specific ways the world might not be alert to possibilities."[41] It's a search for truth, but almost a quiet truth you want to keep a secret. It's about understanding, what does everybody else believe? Why do you think they're wrong? Why are you right?

In a way, independent thinking is a devil's-advocate approach of finding flaws in the currently accepted notions of the deeptech entrepreneurship ecosystem. We need more of this, whether it is for moonshots or otherwise, because we need more breakthroughs, more paradigm shifts, and a lot of exponential progress.

So, be like Anduril founder Palmer Luckey, who is saving soldiers' lives by revolutionizing the controversial field of military technology. Be like Kallyope founder Nancy Thornberry, who discovered a new approach to therapeutics by targeting the gut-brain axis when everyone is chasing the microbiome.

Pursue your contrarian bet—the world needs more of these.

---

41   Eric Weinstein, "The Mind Financing the Future," interview of Josh Wolfe by Astro Teller. *The Portal*, May 3, 2020, audio, 1:42:37.

A GOLDEN AGE
FOR DREAMS

———

"If you want to build a ship, don't drum up
people to collect wood and don't assign them
tasks and work, but rather teach them to
long for the endless immensity of the sea."

—ANTOINE DE SAINT-EXUPERY

OPTIMISM AS A MINDSET

"We have fallen upon evil times, and the world has waxed
very old and wicked. Politics are very corrupt, children are
no longer respectful to their parents."

No, this is not a boomer complaining about Gen Z. Rather,
it's from King Naram-Sin of the Akkadian Empire—which
dates back to around 2250 BC.*,42

———

42 G.T.W. Patrick, "The New Optimism," *The Popular Science
    Monthly* 82, (1913): 492–494.

I mention this quote because University of Iowa professor G.T.W. Patrick wrote about it in a 1913 essay entitled "The New Optimism." Patrick argues there are three kinds of optimism displayed throughout history. First, there is "a priori optimism," which is the extremely outdated notion that whatever the "Creator" made is the best possible thing. The second development is "inductive optimism," or the naive thought that everything in the world is "good and fair" and "full of happiness." Simply put, it's a pretty shallow understanding. The last is "dynamic optimism," which rejects the judgment of whether the world is the best possible one, but rather embodies the motto: "the world is pretty good, but we will make it better."[43]

Patrick recounts the ancient Naram-Sin quote because he argues optimism is killed by people who yearn for the past or the "good ole days." He says, "It usually comes from older people whose outlook may be biased by subjective conditions due to decaying powers and by the tendency to regard all changes as changes for the worse." It's surprising, yet it completely embodies the cultural norm in our society: most people are genuinely pessimistic. Whether it is nostalgia or conservatism (not politically, but being averse to change), some people tend to believe modern times are worse than before and any sense of progression is inherently flawed.[44]

It's crazy to realize the Naram-Sin quote still rings true to this day. In fact, a YouGov poll found "5 percent of Britons think that the world, all things considered, is getting better... [and] 6 percent of [Americans] think that the world is improving. More Americans believe in astrology and reincarnation

---

43 Ibid.

44 Ibid.

than in progress."[45] While the *Spectator* article I found was written in August 2016, I think it's safe to say that number, now in 2020, has decreased for Americans.

Pessimism is a widespread problem. I can't even begin to tell you how many times I've heard someone say, "I'm not being pessimistic, I'm being realistic." Nonetheless, Patrick found "the springs of progress are in the human mind itself," which is exactly why I am writing this section.[46]

Moonshot builders are the people who embody and execute upon dynamic optimism, as they recognize while we are in an age of abundant capabilities, there is still massive work to be done and global problems to be solved. Unsurprisingly, I found founders are some of the most positive out there, as they use it as an energizing perspective shift as opposed to embracing rampant negativity.

To validate this shift, take the sheer rate of scientific and technological advancement: we're constantly pushing the future frontiers at nearly an exponential level. The gold standard: Moore's Law. This states the number of transistors in a microchip doubles every two years, increasing efficiency, while the cost of computers is cut in half in that same time period. This entails an ever-increasing improvement within computing power over time—attesting to futurist Ray Kurzweil's Law of Accelerating Returns.

Kurzweil's law states "technological change is exponential, contrary to the common-sense 'intuitive linear' view."[47] Essentially, he observes, much like Moore's law, deeptech

---

45   Johan Norberg, "Our Golden Age," *The Spectator,* August 20, 2016.

46   G.T.W. Patrick, "The New Optimism," *The Popular Science Monthly* 82, (1913): 492–494.

47   Kurzweil, Ray, "The Law of Accelerating Returns," *Kurzweil Accelerating Intelligence* (blog), *Essays*, March 7, 2001.

fields are progressing exponentially over time. If you consider our breathtaking advances in agriculture, genomics, global infrastructure, computing, and so much more, you will see how we are nearing a golden age with never-before-seen abilities and opportunities within emerging tech.

There are two types of people in this world, optimists and pessimists. It's wrong to equate pessimism to realism, however. That, to me, creates a cycle almost like learned helplessness. Sure, there are times when it is better to look the other way—that is, in fact, realism. Cyclic pessimism means having a mindset that restricts any form of growth, whether it is intellectual, moral, or physical. Some things may be mere obstacles in an uphill journey as opposed to failures that will derail your entire progress. Thus, the key is simply using optimism as a perspective shift. It means recognizing how far we've come while knowing someone has to be the one to continue that momentum.

Letting yourself grow is not about delusion, fantasy, and idealism. It's rather internalizing the confidence that you *can* make that moonshot. Suddenly, you can do anything.

DREAMING BIG

In all honesty, the craziest moonshot I know of is not even a moonshot. It's a starshot.

It's Yuri and Julia Milner's Breakthrough Initiatives organization. It is a collection of scientific and technological explorations into answering the biggest questions in life: what's at the edge of the universe? Are aliens real? Are there habitable worlds? I admire their Starshot project the most. It's a $100 million R&D program aiming to demonstrate technology that enables ultra-light unmanned space flight at 20 percent of the speed of light so we can lay the foundations for an Alpha Centauri flyby mission within a generation.[48]

How cool is that? Just imagine the discoveries that can come from it if it proves to be successful. While Breakthrough Initiatives is more of a nonprofit research project than a business, the core mentality behind it is the same: dream so big people think you're insane.

This is optimism in action. Moonshot operators ask themselves, "Why not more?" and execute upon it with conviction. Why settle for the moon when you can colonize Mars? Why make Band-Aids when you can improve human longevity on a molecular level? Why get better energy output from traditional resources when you can aim for fusion?

One aspect of it is being crazy *is* the way to go now: challenging norms, discovering truths, and exposing inefficiencies are all part of what makes a moonshot founder. It's not about believing in fantasy, but it's about visualizing a far-off solution and channeling an ambitious determination to reach it, whether it be building the world's first commercial space station (Axiom Space) or creating the future of sustainable skincare (Seaspire Skincare). I've even heard of moon hotels, asteroid mining, and underwater cities.

The next aspect is taking a long-term view, or going the extra mile and working on hard issues that might be solved with a five-to-ten-year timeline. The technology or science may or may not be possible within a decade, but regardless, expanding the scope of how to solve a problem inherently makes it more impactful. Take the example of autonomous vehicles (AVs). Talks about these cars have dated back to the '80s, yet only in this century did they really take off. No one

---

48 "Starshot," Breakthrough Initiatives.

knew when the technology would be feasible, as we've been working on it for five decades by now (1980s–2020s). But that did not stop anyone. They dreamt their sci-fi dreams, and only recently did Waymo, Cruise, and Zoox begin to make tangible traction in AVs.

With a next-level optimism and penchant for going bigger, better, faster, and longer, some deeptech founders end up imagining the most unrealistic futures and manage to materialize them regardless.

Unfortunately, the world lacks people with determination. It's the fuel of the inner, mental moonshot that pushes a founder to act upon their dream. We need to inspire people to imagine more. "The limit on global progress is the ambition of the most ambitious," remarks Navid Nathoo, co-founder of The Knowledge Society—a school for "moonshot" students. Just imagine what would happen if everyone dreams just a little bit bigger. The possibilities are endless.

TOMORROWLAND

While the future may be coming faster than we think, it won't do so if we just wait for it. It doesn't happen to us, but rather because of us.

That's why moonshot founders think like futurists, who are people who attempt to explore predictions and possibilities about what's to come. In other words, they dream big in a more pragmatic way. They extrapolate the potential patterns and how we will get there based on elements of reality and how we are progressing as a whole. By now, I've mentioned Alan Kay's famous quote many times. Yet, if "the best way to predict the future is to create it," then putting a futurist perspective behind by a moonshot mindset allows one to envision a world in which a problem is no longer there and work backward from it to figure out the intricacies of the solution.[49]

To garner actionable foresight, start by considering the sub-trends, variables, and externalities leading toward four key futures: the projected future, plausible future, possible future, and preposterous future. Respectively, these represent the full range of potential events, or what will probably happen, what might happen, what can happen, and what is the most extreme thing that can happen. Then, within those, one must find their preferred future.[50] Being in deeptech puts anyone in the right position to do this, as they are already on the cutting-edge anyway.

I've observed two ways of doing that. The first is by finding inspiration and value in the most unexpected places. In *Zero to One*, author Peter Thiel recounts "the single most powerful pattern [he] has noticed is that successful people find value in unexpected places."[51] In other words, they stay optimistic and ambitious about everything that can point to the end goal they seek, no matter how seemingly insignificant it is.

A perfect example of this is Bolt Threads. Founders Dan Widmaier and David Breslauer created the sustainable biofabrics company with the goal to create clothing materials from natural sources such as mycelium (mushrooms) and silk proteins. The key consideration is Widmaier and Breslauer

49 Alan Kay, "Early Meeting in 1971 of Parc, Palo Alto Research Center, Folks and the Xerox Planners," (meeting, Xerox Palo Alto Research Center, Palo Alto, CA, 1971).

50 Lisa Kay Solomon, "How Leaders Dream Boldly to Bring New Futures to Life," *SingularityHub,* February 23, 2017.

51 Peter Theil, *Zero to One: Notes on Startups, or How to Build the Future* (New York, NY: Currency, 2014), Preface.

were inspired to make such a company by watching how spiders created silk. They've fundamentally bioengineered the natural process on a larger scale for sustainable materials with incredible benefits such as increased strength, elasticity, durability, and softness. Clearly, they found value in spiders and the forest floor.[52] It's quite unexpected in my opinion, but it just goes to show moonshot founders can reach their preferred future with a strategy completely out of left field.

The other way in which one can adopt a futurist mindset and let their imagination go crazy is through backcasting. Mike Maples, co-founder and partner at Floodgate, found this technique is a legitimate way to make a breakthrough. He recounts how Steve Jobs did not "discover" the market for smartphones, but rather designed the industry and taught us how to appreciate it. Elon Musk, Jeff Bezos, and Richard Branson did the same with commercial space travel—they singlehandedly carved out the industry rather than simply identifying it. As such, Maples describes these legendary founders as "breakthrough builders [who] are visitors from the future, telling us what's coming."[53]

Essentially, this means no matter how preposterous they sound, these founders are the ones who bring the current reality into their preferred and designed future. This is called backcasting. It's different than forecasting, which views the world in an incremental trajectory and is inherently flawed due to the fact it applies present assumptions to the future.

52 Hello Tomorrow, "In the Future We'll Wear Spider Silks | Dan Widmaier | HT Summit 2017," Hello Tomorrow, November 20, 2017, YouTube video, 15:59.

53 Mike Maples Jr., "How to Build a Breakthrough," Mike Maples Jr. on Medium (blog), *Medium,* April 27, 2020.

To be a futurist, one must think about what the products, markets, and ecosystems will look like and then trace backward what happens in the process of getting there. Maples posits this is done by first looking for inflection points, or moments in which the standard cadence of advancement changes drastically.[54] In other words, look for a sharp trend shift in:

- Technology, or platform technologies such as the internet, AI, and other exponential improvements.

- Adoption, or forces that can enable incredible demand for another product (how smartphone adoption boosted Uber, Lyft, etc.).

- Regulatory, or changes on a public policy level that are conducive to the given breakthrough.

- Belief, or the underlying believability in a certain deep-tech field.[55]

What this all means is by watching out for inflection points, you can find opportunities to capitalize on them before they become widespread and thus use that to build a breakthrough. Once you reason out the inflection points that will lead to the preferred future, whether it is possible or preposterous, you must gather insights around it and bet on yourself because you're in for one hell of a ride.[56]

---

54 Ibid.

55 Ibid.

56 Ibid.

Ultimately, moonshot founders are those who combine optimism, imagination, and futurist thinking. Together, this has the power to accelerate the course of scientific and technological progress for all of humanity. Yet, I must return to my first point and concede that not everyone has the ability to dream big. Not everyone is safer, healthier, or prosperous enough to take intellectual and entrepreneurial risks.

That is why I said we're so close to a golden age. We need to build for the betterment of all instead of allowing the polarization of progress in deeptech.

To do that, we must spark a movement. A renaissance, per se.

---

*   In Professor G.T.W. Patrick's essay, which was written in 1913, it is mentioned the quote was written by King Naram-Sin of Chaldea in 3800 BC. He says he found the quote inscribed on an old stone in a museum in Constantinople. This was before the advent of carbon dating (1946), so his estimation was most likely incorrect. King Naram-Sin was not the ruler of the Chaldean Empire, which was started around 626 BC. He was the ruler of the Akkadian Empire in the years 2254 to 2218 BC.

AUDACITY CAN BE EASY

"The future is open. It is not predetermined and thus cannot be predicted—except by accident. The possibilities that lie in the future are infinite. When I say, 'It is our duty to remain optimists,' this includes not only the openness of the future but also that which all of us contribute to it by everything we do: we are all responsible for what the future holds in store. Thus, it is our duty, not to prophesy evil, but, rather, to fight for a better world."

—KARL POPPER

ON DARING TO ADVANCE HUMANITY

It's one thing to build a company. It's another to establish a city.

Charter cities are jurisdictions granted the special ability to create a new governance system and pick the best practices regarding everything from economic development to laws. It's probably one of the boldest initiatives I've heard of.

Think about it. This means generating two core aspects from scratch: supply and demand. Planners must determine how to build infrastructure, hospitals, housing, central banking systems, commerce, laws, law enforcement, governing structures, education, jobs, sanitation, and other basic human necessities. Not to mention, you have to consider real estate development, international geopolitics, macro- and micro-economics, livability, innovation, policies, and so much more. Charter city organizations seek to build entire ecosystems, entire civilizations.[57]

There's something so philosophical about charter cities. It takes from ancient wisdom in how to make civilizations, how to unlock human potential, and how to ensure a virtuous existence. It's influenced by what worked throughout all of history. On the one hand, these cities enable socioeconomic mobility and spark golden ages of art, science, and technology. On the other, they boost moral progress and prosperity for all. Imagine being able to recreate a modern-day Renaissance.

The case for these moonshot cities runs parallel to that for moonshot companies: doing big things to solve big problems. While they are fundamentally impactful and can improve the livelihoods of many, there are not enough out there to supercharge a movement for them. The few that exist, however, are incredibly successful: nearly $375 billion GDP in Shenzhen,[58] $372 billion GDP in Singapore,[59] and $366 billion GDP for Hong Kong.[60]

---

57   Charter Cities Institute, "An Introduction to Charter Cities."

58   Leng, "China's Tech Hub Shenzhen Set to Hit 2019 Growth Target."

59   Trading Economics, "Singapore GDP."

60   Trading Economics, "Hong Kong GDP."

We don't have more of these despite their extremely evident upside because of who we are. We've incentivized the wrong things, such as corporate greed and a toxic individualism, and stymied the rate of genuine improvement. That, in turn, means fighting innovation, as it inherently brings change and power shifts. It entails helping oneself at the expense of others. Human nature is our own worst enemy.

I sat down with Anirudh Pai, a researcher at Pronomos Capital, a firm that invests in and advises charter cities. Upon asking how to make more of these moonshots, he said we must navigate the physical and the philosophical aspects by initiating a widespread movement toward progress and a better future. If we can get people to collectively agree that radical improvement is a good thing (shocking, I know), then we can make entire city-states.

I've heard that notion many times. As a society, some simply do not believe in progress. There needs to be something as widespread as a religion behind building for the better. Charter cities are the embodiment of that idea. They give people the power of choice, autonomy, and a newfound potential. Maybe consider it that they grant life, liberty, and the pursuit of happiness for all. We need a culture of doing bold things. That's the key to it all.

It's overwhelming to start such a massive project considering every single aspect of life itself. I asked Pai about this, and he recounted the quote, "How do you eat an elephant? One bite at a time." This is by no means an easy task. It begs the question, who dares to do such a thing?

Moonshot founders.

CLEAR VISIONS AND ANTIFRAGILITY

Being audacious is about the execution as you act upon lofty goals.

It's not about being blatantly unrealistic, nor extremely rash, nor absolutely fearless—although you need some of those qualities to a lesser degree. Instead, it's about creating the largest impact humanly possible for yourself. It's about going the extra mile when anxiety and uncertainty are discouraging you.

In that light, it's useless to write, "Oh, just be more audacious!" That adds no value. Thus, the perspective shift and appropriate behaviors are as follows. You need the mental foundation for audacity. Without it, a founder is just plain reckless. This means putting conviction behind a clear vision, which gives it context and augments the drive to achieve it.

That said, I came across the answer in a conversation with Arvind Gupta, co-founder of IndieBio and partner at Mayfield. He's a leader in the moonshot biotech space, so I asked him how he invigorates the creative juices and courageousness within the companies he accelerates.

Gupta recounted a talk he has with every founder. Essentially, he recommends they "reframe" their lives around the change they wish to create. In other words, they should have self-awareness and intellectual honesty to realize why they genuinely want to go about making the moonshot. The impact they seek to create must be their North Star, and they must align it with their life's purpose. For the success stories within his portfolio, the common thread was founders did just that.

It's a simple yet powerful tool because this is where the value lies. Audacity doesn't mean moving fast and breaking things, as Silicon Valley loves to say. It means diligently and systemically going the extra mile and willingly taking risks that align with the original vision.

But, that's not it entirely. As courageous as a vision is, the uncertainty of entrepreneurship can diminish it in a split second. Thus, a major part of being more audacious is antifragility. This does not mean being resilient, persistent, or thick-skinned. The word antifragile may be thrown around and used in various contexts, but the specific definition relevant to moonshots is from Nassim Taleb. An expert in randomness, probability, and uncertainty, Taleb describes this quality in his book *Antifragile: Things That Gain from Disorder* as "[benefiting] from shocks," "[thriving and growing] when exposed to volatility, randomness, disorder, and stressors," and "[loving] adventure, risk, and uncertainty." Simply put, it means becoming stronger under pressure and uncertainty. He says that "antifragility has a singular property of allowing us to deal with the unknown, to do things without understanding them—and do them well."[61] While this concept is typically used to describe systems, such as political structures, economic markets, revolutions, and even literature, it is also seen in the moonshot founders. After all, the best inventors are patient and steadfast despite the adversity they face.

This is because of how complex the nature of their work is. Aiming for an even higher impact imposes more stressors and exponentially greater risks, so the founder must be able to withstand the headwinds that come with their efforts. Being bold increases the level of confusion and unpredictability, but when you are obsessively committed to precision yet are able to thrive in a do-or-die atmosphere, you know you can power through anything.

---

61 Taleb, *Antifragile*, 29–31.

BHAGS: BIG, HAIRY, AUDACIOUS GOALS

If clear visions mean relentlessly staying in your lane, then BHAGs mean going at supersonic speed.

I'm all for dreaming as large as you want, but there has to be a reliable path to getting there. Therefore, inspiring long-term goals need intelligent medium-term milestones and practical short-term guideposts. Because the subgoals must be achievable, one recommendation is to do the hardest thing first. It seems counterintuitive, but the risk-reward equation is ever-more prevalent the younger a company is. Trial by fire will determine if the company has a future because time is the primary constraint, not money.

Also, to combat the complexity, it is recommended to distinguish scientific-technical milestones and business milestones. For example, the BHAG for charter cities is to create a near-utopian, prosperous society. But to get there, you need to take baby steps in discerning which goals to achieve first, whether they are infrastructure and real estate, capital and commercial flows, or a governing structure. For deeptech startups, that means keeping R&D goals separate from business ones so you can be tenacious within each realm without overwhelming one another. Create synergy to lessen the chaos.

This results in a culture in which people are individually creative but collectively reach for BHAGs. No one will be ridden with self-doubt, and everyone will be empowered to do things they previously believed they couldn't. To do that, companies should seek to make audacity the path of least resistance. Allow yourself and team members to feel psychologically safe in pushing against comfort zones. Brainstorm foolishly, because unobstructed thinking leads to better potential routes. Know you cannot get to the extraordinary ideas without spending much time building upon the bad ones.[62]

In my conversation with Arvind Gupta, he specifically stated some companies are not as bold as they can be. He noted the case of Girihlet, a biotech startup that developed a way to sequence T cell receptors. At first, Girihlet simply sold sequencing services, which was fine as it is. But as they did the work, they realized they had a good diagnostic tool that could identify autoimmune diseases. Then, they realized they could develop antibodies and therapeutics against those targeted diseases. And thus, by having a clear vision and letting the company be bolder, Girihlet transformed from sequencing to curing autoimmune diseases.[63]

As more and more moonshot companies aim for greatness, it is vital to recognize society's role in being bold. Cultural influences, societal structures, biases, heuristics, stereotypes, and more are all stopping founders from being much more daring, which blocks these companies from fulfilling their highest BHAGs.

The best way to fix that on a societal level is to make more of them. Ideally, successful ones. Once you ignite radical creativity, zealous enthusiasm, and heroic bravery within a team, the energy is contagious—"10x can light a fire in people's hearts in a way that 10% can never do."[64] This dynamic also bleeds into other startups. Making one moonshot will inspire another, which will inspire two more. First, we have SpaceX aiming for easier spaceflight and Mars colonization,

---

62 Astro Teller, "The Head of 'X' Explains How to Make Audacity the Path of Least Resistance," *Wired,* April 15, 2016.

63 Girihlet, "What We Do."

64 Gimbal_v2.0_X (PDF), X Moonshot Factory, accessed August 30, 2020.

then we have Blue Origin, Virgin Galactic, Rocket Lab, and many more spacetech firms. First, we have Eat Just developing sustainable cell-grown meat, then we have Finless Foods and Memphis Meats. It's all a "continuum," as Gupta calls it: a continuum of crazy vision, execution, and impact.

We've lost our innovative touch and our penchant for doing crazy projects. The way I think of it—if someone can eventually recreate what you are working on, you aren't ambitious enough.

Human emotion is the biggest obstacle toward audacity, so if you can tackle that, society will get what it ultimately needs.

TWO IS BETTER
THAN ONE

"Invention, it must be humbly admitted, does not
consist in creating out of void but out of chaos."

—MARY SHELLEY

VARIED BACKGROUNDS

As much as deeptech is science, it's also an art.

In a way, it matches Pablo Picasso's approach to painting. In the first decade of the 1900s, Picasso underwent rapid experimentation and began questioning everything he ever knew. He was dissatisfied. He wanted to break the rules. For years, Picasso painted in a more emotional and representational manner, but he flipped the script when he took a step back. Instead of the inner, he questioned the outer: space, existence, vantage points. He challenged Renaissance depictions of rigid and modeled forms and instead opted for dynamic arrangements of his works' background and foreground. As he stepped out of the boundaries of what was generally accepted, he ended up sparking one of the most trailblazing movements of the 1900s: Cubism.[65]

In a completely different realm, Albert Einstein wholly altered our understanding of physics with his newfound theory of relativity. Sitting in a patent office, he posited relativity can be applied to gravity, which occurs from the warping of space and time between certain masses. It was the same structure as any scientific revolution. First, something wasn't right. Then there was a considerable unraveling. Einstein ended up changing our knowledge of the universe as a whole.

I mention these stories because there is a myth that says Pablo Picasso and Albert Einstein met in a Paris bar one day in 1905, revolutionizing art and science, respectively. While it is almost certain this event didn't happen, there is a clear link between Cubism and the theory of relativity: both have to do with the nature of space, perception, and the fourth dimension. Whether it is real or not, the story portrays the value of diversity, perspectives, and combination—especially for breathtaking moments and cultural shifts. It's not every day someone completely changes the course of humanity in the arts and sciences as they did.

It seems pretty obvious that having fresh perspectives is essential to entrepreneurship as well. It doesn't take an expert to realize that. But I found moonshot companies take diversity to the next level: from background to thought to innovation.

Everyone *should* know the importance of diverse people in a business setting. It's sad some don't value it. But from the dawn of business books and management guides, every author has stated its significance. It's a saturated concept now.

---

65 History.com Editors, "Cubism History."

Moonshot companies are made by diverse teams of people who each individually add value to the organization. The surface level of this is conventional diversity, which entails race, gender, sexual orientation, nationality, and more. It's also important to note operators may also come from varied social classes. Countless success stories of those who are self-made and came from a poor background exist. These people tend to be "PhDs": poor, hungry, and driven. Just as many people got rich and made the impact they wanted to. The epitome of this is Bill Gates, who started Microsoft out of a garage and has since spent over $50 billion in charities and research (Malaria cures, vaccines, and the like).[66] Finally, some are inspired due to personal tragedy; others are motivated by the crumbling world around us. At the root of all this is one thing: differences in background.

Nonetheless, an evident pattern within deeptech startups is they purposefully seek people with skills in diverse sciences and technologies—genomics and engineering, materials science and biology, neuroscience and computation. This is the reality at the X Moonshot Factory, which strategically hires extremely diverse experts for their projects. For example, the team behind Everyday Robot, which is trying to build general-purpose robots for an unstructured world, is comprised of machine learning PhDs, puppeteers, Mars rover engineers, Marine helicopter mechanics, and even a chocolate maker. While the main purpose of the effort is for robotics, X implements experts in deep learning, shared experience, and human demonstration—hence the puppeteers—to find new solutions and possibilities from a rich diversity of professions.[67]

---

66 Brown, "Bill Gates $50 Billion to Charitable Causes."

67 Hans Peter Brondmo, "Robotics Is a Team Sport," *X, The Moonshot Factory* (blog), *Medium,* June 5, 2019.

In other terms, moonshot companies are typically run by teams and cultures of cognitive diversity, on top of those who have unique surface-level backgrounds. With the context, combination, and compounding of everyone's expertise comes an augmented sense of radical creativity built upon knowledge both wide and deep.

By taking a more in-depth look as to why most companies do not emphasize this, it's clear it is not necessarily helpful for every team. In terms of skills, whether it is for investment banking, software engineering, or any other job function, organizations don't look for people who are unskilled within the field at hand. I've never seen McKinsey adding astronauts into their management consultancy.

Colloquially, it's called the Stormtrooper problem. When founding teams recruit multiple people who are pigeonholed within one field and only let those people work on problems solvable with their skillset, you end up hiring Stormtroopers who are all the same and are cut from the same cloth. For example, the software engineers stick with the software engineers and the growth marketers stick with the growth marketers, but there is little cross-play between the groups. Everyone looks at a problem the same way, and there is no one checking their way of thinking. I'm not saying that strategy is wrong. Most unicorn companies probably do just that and are incredibly successful.[68]

It's simple. You do not typically hire people who do not have the skills or background for a certain task. But the thing is that moonshot companies do the exact opposite. While traditional companies seek differences in social terms (race,

---

68 "The Stormtrooper Problem: Why Thought Diversity Makes Us Better," *Farnam Street, fs.blog,* accessed August 26, 2020.

gender, etc.) and business terms (perspectives, skills), early stage deeptech firms aim for those things plus differences in intellect and cognition. Essentially, moonshots tend to have cross-collaboration between subject matter experts who can take apart the complexity of a problem and solve it on each other's backs. The value comes from their differing mental frameworks clashing together to fill the gaps of each other.

It's all about explaining a solution to a problem when one field of study cannot—for example, genetics and anthropology for drug discovery (Variant Bio) or fashion designing and aerospace engineering for durable stratospheric internet balloons (Loon).

COMBINATORIAL INNOVATION

Diversity of thought leads to interdisciplinary thinking. This entails many things, such as a deeper system of checks and balances on others' thought processes or combining the knowledge base of various domains to uncover interlinked value. In other words, the emergence of patterns and regularities from the aggregation of many scientific and technical fields.

If diversity is key toward a species' survival, a diversity of thought enables longevity. Likewise, the former is integral for a startup to merely stay afloat in the world of entrepreneurial Darwinism, but collective intelligence allows a team to truly go far.

However, that's not the important part. With all the interdisciplinary energy, the most impactful groups make it safe to speak your mind and self-radicalize. Members go further outside of their niche as they explore ideas and actions they otherwise would not have encountered. This requires a culture of trust before anything else.

Homogeneity of vision can be a good trust-builder, since you can diverge in opinion without losing the sense that you are working toward a common goal. A vision is less powerful if it's imposed, yet much more powerful if it results from a group process: jamming, contradicting, exploring. There need to be some shared goals, norms, and context.

To back that up, we can look to countless instances in which a bunch of smart people with unique identities came together and had transformative results. Take Ernest Shackleton's crew in his expedition to Antarctica. He had navigators, sailors, biologists, meteorologists, carpenters, geologists, motor experts, physicists, and many more people. With Shackleton's outstanding leadership and the sheer excitement of his crew to reach the bottom of the Earth, the diverse team did something only a handful of people would ever attempt to do. It was all because of the diversity of thought, not to mention the mutual perseverance that came from it.[69]

Take the Manhattan Project. Although controversial, it's an example of people from many different sciences coming together and producing something world-changing with physics, chemistry, military tech, and more. Take Bletchley Park, another World War II example in which the smartest minds cracked the German Enigma machine with cryptography and mathematics. You can consider so many things: the Apollo missions, the Bauhaus, the X Club, Xerox PARC, Bell Labs, and now countless companies, from Apple (computers and design) to Celevity (dogs and molecular biology) to Vardaspace (space and manufacturing).

---

69 "Ernest Shackleton's Crew of the Endurance Imperial Trans Antarctica Expedition 1914 -17," Cool Antarctica, accessed August 26, 2020.

It just goes to show if you take a concentrated area of smart and intellectually diverse people, you will get groundbreaking results. You will get moonshots.

In the end, the value of all of these perspectives is you get combinatorial innovation. This concept means using the deeptech building blocks and coming to discoveries and capabilities as a result of the novel, combined insights of the fields. For places where conventional wisdom may not be enough to tackle problems such as catalyzing clean energy or augmenting crop yields, combinatorial innovation steps up.

Moonshots do not occur in a vacuum, but rather at the crossroads of various ideas, backgrounds, skills, thought processes, and cognition. The compounding knowledge breeds radical creativity and drives human, technological, and scientific progress forward.[70] With every step-change and advance, it is clear that being interdisciplinary is the key.

Like that, the collection of fresh perspectives and synergistic community wisdom propels forth the rough ideas and uncomfortable moments necessary to solve pretty challenging problems.[71] And when a startup carves out a "safe space" for the weird souls—much like what Willy Wonka did for his Oompa Loompas—the multifaceted inventive spirit is amplified to outrageously radical extents.

---

70   Steve Jurvetson, "Moore Evermore in Computer History
     — Happy 50th Birthday to the Law!" *Steve Jurvetson
     photostream* (blog), *Flickr,* February 11, 2015.

71   Astro Teller, "Tips for Unleashing Radical Creativity," *X, The
     Moonshot Factory* (blog), *Medium,* February 12, 2020.

3

PHILOSOPHY

MORE CELESTIAL
POWER TO YOU

WE'RE ALL DEAD STARS.

That's right. NASA astronomer Dr. Michelle Thaller theorizes that humans are the remains of the supernovae of the distant past. A supernova is a Herculean, luminous explosion that occurs when a star nears its final evolutionary stages. It bursts into a vibrant show of elements such as hydrogen and helium and transcends great distances. With time and the perfect environmental conditions, the atomic remnants led to the creation of nearly all celestial bodies. That means hundreds of billions of dead stars contributed to the advent of life on Earth, including us.[72]

It's empowering, to say the least. It seems like we're supposed to be among the stars, doing things worthy of these gargantuan astronomical events that fundamentally made us.

---

72 Michelle Thaller, "We Are Dead Stars: We Are Born of Supernovas – Our Spectacular and Totally Ordinary Origin Story," September 14, 2017, in Aeon, produced by Laura Lichtman and Katherine Wells, video, 3:56.

That's exactly what moonshot companies do. They are individual supernovae that augment our existence, uncover basic truths, and ease human suffering. We all have the astral spirit within us. It's just a matter of how we harness it.

In doing so, I realized emerging tech operators not only tend to behave a certain way, but they also view the world differently. There are an infinite number of philosophies out there, yet those employed by people on the cutting-edge point to what they consider, how they process it, and how they translate it to their work. Let's delve into the key inner dialogues, thought processes, and mental models. I will tell you about the connection between humanity and moonshot startups and how to interpret the process of developing one.

Understanding the self is the first step in helping others and doing impossible things. Once you master the self, you can master the moonshot and supercharge the positives that arise with it. We must teach everyone how to *view* building big things. If we wholeheartedly believe progress is desirable, making an impact is our life's purpose, and the bleeding-edge unlocks novel solutions, then we can make more moonshots.

At the end of the day, we *can* be amazing.

THE MORAL IMPERATIVE FOR MOONSHOTS

"But without scientific progress no amount
of achievement in other directions can
insure our health, prosperity, and security
as a nation in the modern world."

—VANNEVAR BUSH

HUMANITY'S LOWEST POINT

Austria, 1945. Neurologist, psychiatrist, and Holocaust sur-
vivor Victor Frankl returned home to Vienna from the Aus-
chwitz concentration camp. As he recounted the treacherous
experience, he realized the camp's sheer brutality brought to
light an important truth about human struggle.

After enduring the excruciating terrorism, Frankl
wrote about that truth in *Man's Search for Meaning,* which

chronicled the psychology behind the survivors versus those who unfortunately passed away. He found that the people who had some ultimate duty waiting for them upon liberation, despite their abominable circumstances, were those who survived. These were the people who had an irrepressible feeling of responsibility—a purpose—they had to fulfill in the future: love, family, lessons learned, life itself, anything. Frankl found "man's search for meaning is the primary motivation in his life," and those who gave up on that search, those who were so ingrained in their past, had no will to live and died because of it.[73]

Frankl writes, "This meaning is unique and specific in that it must and can be fulfilled by him alone; only then does it achieve a significance which will satisfy his own will to meaning."[74] This psychotherapeutic concept is called "logotherapy." It's the notion that our prime motivation in the world is to find our purpose, whether it is through what we do, what we experience, or what we stand for.[75]

He also proclaims we can discover the meaning of our lives in three ways: "creating a work or doing a deed," "experiencing something or encountering someone," or "the attitude we take toward unavoidable suffering."[76] The common theme among these is we must view life as a positive-sum game: not zero-sum, and not negative-sum—although humanity's inherently self-serving nature may make people believe that.[77] Likewise, everyone is irreplaceable in their

---

73 Frankl, Man's Search for Meaning, 104.

74 Frankl, Man's Search for Meaning, 106.

75 Viktor Frankl Institute of Logotherapy, "Logotherapy."

76 Ibid.

77 Frankl, Man's Search for Meaning, 116.

mind, body, and spirit and can individually better society in our own way.

That said, many are misguided to varying degrees. Victor Frankl devised this theory by observing humankind at one of the rawest, most vulnerable moments in history. It just goes to show anyone can truly adopt this philosophy.

OUR COLLECTIVE POTENTIAL

If we're all here on this planet to achieve our human potential, then let's align that with building things for the greater good. Out of many paths to do so comes entrepreneurship, mainly because it is a way for private individuals to make the impact they seek of their own accord. Take it one step further, and we have moonshots, which are the epitome of maximizing this potential.

If you think about logotherapy as it relates to deeptech, it means you have a higher chance of succeeding if it is a natural extension of yourself or your purpose.

Why? Because we *all* need you to do it. Only you.

Imagine the loss society would incur if you do not pursue your ultimate aspirations. What if you do not chase your greatest purpose, your meaning of life? In a way, you're hindering the possible greatness that can grace the world. Your inaction is why some future course of events will not happen for any arbitrary individual's betterment.

Instead of viewing it like "Should I pursue X, Y, or Z?" it is instead "What will happen to the people I could've helped had I not given up?" I believe this notion presents an imperative that only you can make that moonshot you think the world needs. Let that sink in.

THE NECESSITY FOR TECHNOLOGY

I'm not the only one who believes this, however. Take Josh Wolfe, co-founder and managing partner of Lux Capital—an emerging-tech venture firm. Wolfe believes there is a moral imperative for people to invent technology to express their genius and ultimately become more human. This essentially means reaching one's utmost capabilities because it is inherently beneficial for at least one other person in the world.

He posits, "Imagine a world in which Person X existed and Technology Y didn't. Imagine a world in which Mozart exists, but the harpsichord doesn't. In which Hendrix exists, but the electric guitar doesn't. In which Spielberg exists, but the eight-millimeter-camera doesn't. [In which] Bill Gates exists, but the PC doesn't."[78]

The world needed those individuals to fulfill their purpose. Just think about how much their actions reverberated throughout history: how many people they inspired, how grand of a golden age they sparked within their industries. Then imagine them not doing it. What would their beneficiaries' lives be without it?

Nonetheless, that backward hypothetical thinking sets the record straight: you have an imperative to attempt that moonshot because without trying, you will never materialize the better future that could've been.

To add to that, imagine a world where John F. Kennedy did not push for reaching the moon. Or a world in which Martin Luther King Jr. and Civil Rights leaders did not galvanize the masses to push for equality. Or even a world in which healthcare workers did not sacrifice their lives to treat

---

78 James O'Shaughnessy, "Josh Wolfe – The Tech Imperative," April 23, 2019, in Invest Like the Best, produced by The Investor's Field Guide, podcast, MP3 audio, 1:17:15.

COVID-19 patients. The imperative to do what's difficult transcends deeptech. Their radical creativity and pure conviction—in tech, politics, medicine, or otherwise—had an impact so profound that had they not channeled it, the world would not be the way it is today. People would not be as better off as they are at this moment.

As Wolfe put it, "Every one of these things was an instrument for them to express a form of genius, and there is somebody out there now that is going to encounter technology in ten years that doesn't exist today that they will play as their instrument to the world, and the world will be better off because of it."[79] That person is you. It's up to you to invent it.

Again, everyone has a moonshot in them. No matter the platform of expressing your genius, there is something out there for everyone. Everyone has their own opinions, backgrounds, and passions. They must take the initiative to achieve their innate genius because it will positively benefit someone somewhere.

Wolfe encapsulates my thoughts with his call to action: "If you could do two moral things, hit that moral imperative to invent so people can find their genius and then find ways to reduce human suffering."[80] If our primary motivation is to find our purpose, and our purpose is to solve the tough problems we face, then we will do it so long as it's humanly possible.

Some say we must create more technology. I say there is a moral imperative to take moonshots, mainly because a world in which we don't do the hard-but-necessary things is a recipe for a dark future.

---

79 Ibid.

80 Ibid.

MASTER THE MIND TO MASTER THE WORLD

"In one soul, in your soul, there are
resources for the world."

—RALPH WALDO EMERSON

VIRTUE

Many people believe technology and philosophy are polar
opposites. They say one is technical while the other is artistic.
One is tangible, while the other is intangible. I would say they
are quite similar in concept. People within each domain seek
to better understand how the world works. They increase our
humanity. They find fundamental truths. Some even believe
"if the proximate purpose of technology is to reduce scarcity,
the ultimate purpose of technology is to eliminate mortality."[81]

---

81   Balaji S. Srinivasan, "The Purpose of Technology," Balaji S.
     Srinivasan (blog), Ghost, July 19, 2020.

Physical and symbolic immortality. That's some deep stuff.

Yet, in terms of philosophy and its expansive history, there are a few key elements that moonshot founders embody. The ancient Greek philosophy "eudaimonia" is one of them. In Aristotle's words:

> *"People of superior refinement say that it is happiness, and identify living well and doing well with being happy; but with regard to what happiness is they differ, and the many do not give the same account as the wise. For the former think it is some plain and obvious thing, like pleasure, wealth, or honour; they differ, however, from one another—and often even the same man identifies it with different things, with health when he is ill, with wealth when he is poor; but, conscious of their ignorance, they admire those who proclaim some great ideal that is above their comprehension. Now some thought that apart from these many goods there is another which is self-subsistent and causes the goodness of all these as well."*[82]

In simple terms, eudaimonia entails "human flourishing or prosperity" and "blessedness." It does not directly mean happiness and being content. It is not a state of mind or a dopamine hit that spurs pleasure. It is rather about achieving the highest human good by doing noble things: not just chasing excellence, but also leading a virtuous life that is objectively worth living. In turn, happiness is a byproduct.[83]

---

82 Aristotle, Nichomacean Ethics, Book I, Chapter 4, trans. W.D. Ross.

83 Positive Psychology, "Eudaimonia."

To me, it is about reaching one's potential, no matter what it may be.

In the context of this book, I mention eudaimonia for two reasons. The first is the concept has no absolute English translation. It is a Greek term said to have inspired the meaning of "happiness," but according to Aristotle, it isn't directly that. It begs the question, how can English speakers achieve this empowered state when we do not know of it?

It's like telling someone to think of a new color. Try it. You can only think of shades and combinations of existing ones. Accordingly, the construct of eudaimonia presents a notion on how powerful it is to simply know more. Only then can we adopt it.

The second is something a lot deeper. Every moonshot founder I ever spoke to or researched is inadvertently aiming for eudaimonia. They are dissatisfied with something and want to see it fixed, so they establish a company. Yet, the journey is what makes them feel blessed.

Jonathan Tan, co-founder of Coreshell Technologies, said it best. "Not very many will get to step away from their careers and be like, 'Okay, I'm going to start this wholly unknown endeavor with very little chance of success and try to create something and test it out.' I realize that that's a privilege in itself and it's something that I don't take lightly. That's what's really been motivating me and keeping me going." For context, Tan is leading the transition to advanced power sources by creating a nanolayer coating technology to stop degradation in rechargeable batteries. He has the drive to solve the technically difficult problem because he has that inner well-being. Doing so makes him genuinely happy.

SURVIVING THE JOURNEY

Unfortunately, that's the hardest thing to do. Reaching that heightened state is no easy task, so it takes a mentally fortified person to do so. This is where stoicism reigns supreme.

Stoic philosophy is not what the English language defines as "stoic." It's not about being emotionless. Rather, stoicism is advice for endurance, courage, temperance, and wisdom. It's a lot, but it is encompassed by the idea of facing obstacles and being positive despite them: how to live the best life in a world full of misfortune.[84]

The biggest part most philosophers agree upon is the best thing to do is to realize what you can control and what you cannot: internal versus external. Do not worry about what you cannot control: others, property, reputation, command, nature, and more. Rather, focus on internal capabilities: opinion, pursuit, desire, aversion, and your own actions.[85]

As a result of internal or external classification, people should channel their ambition, motivation, and energy into what they can influence while mentally detaching themselves from things they cannot. This dichotomy of perception applies to every human out there.

Embracing stoicism also entails embracing misfortune and thriving despite it. It can actually be used for good— assuming one engages in a healthy amount of self-examination, self-reflection, and self-awareness.

This is especially important in startups, as misfortune is an everyday thing. In celebrity entrepreneur Tim Ferriss' blog, it was written:

---

84 Encyclopedia Britannica Online, Academic ed s.v. "Ancient Stoicism," accessed September 11, 2020.

85 Epictetus, The Enchiridion, trans. Elizabeth Carter.

"The Stoics were writing honestly, often self-critically, about how they could become better people, be happier, and deal with the problems they faced. As an entrepreneur you can see how practicing misfortune makes you stronger in the face of adversity; how flipping an obstacle upside down turns problems into opportunities; and how remembering how small you are keeps your ego manageable and in perspective. Ultimately, that's what Stoicism is about. It's not some systematic discussion of why or how the world exists. It is a series of reminders, tips and aids for living a good life. Stoicism, as Marcus [Aurelius] reminds himself, is not some grand Instructor but a balm, a soothing ointment to an injury wherever we might have one."[86]

Newsflash: making a global impact is not painless. You may have a clear path to do so, but then you have to consider every aspect of business, from human resources to sales, politics and regulatory hurdles, and the societal shockwaves that come from it. Trouble is inevitable when founding a company. Take it from literally any entrepreneur out there.

Nonetheless, recognizing setbacks as teachable moments is what builds endurance.

DOING THE RIGHT THING
The path to eudaimonia is unpredictable, but its benefits—personal flourishing and a positive impact—far outweigh the tough journey. While I interpret stoicism as the perspective

---

86 Tim Ferriss, "Stoicism 101: A Practical Guide for Entrepreneurs," The Tim Ferriss Show (blog), tim.blog, April 13, 2009.

shift behind attaining it, I believe the context of the execution is just as important.

This was inspired by my conversation with Jude Gomila. As the founder of Golden, he is building a knowledge engine to compile *everything* humankind has ever discovered: essentially an "order of magnitude better version of Wikipedia." Not to mention, he's invested in more than 200 of the boldest companies out there.

When I asked him how he navigates the philosophical aspects of founding a company versus the tangible side, Gomila spoke of this philosophy of executing on the right things. It's simple. All he said was, "Is that the right moonshot to try and pull off right now?" Is it necessary for us? Is it too early to materialize? Is it too commoditized? Do we fundamentally need supersonic jets or 3-D printing for the moon? Why? Will it create unemployment? Will it ruin markets? Will it create more problems? Is there a return on investment? Or is it just a public relations stunt? Is it hedonistic? Is it worth the time?

The dynamics of his "moonshot selection criteria" are exactly what the context is for the road to leading a virtuous deeptech startup—especially in a world of finite resources.

It's similar to the concept of effective altruism. Sure, we may all seek to make a positive impact, but are our efforts useful? Some social programs don't work, some charities are frauds, and some companies are using claims of corporate social responsibility as marketing tactics.[87]

If people want to "make a difference," they must ask themselves if their method is truly productive. Especially in the realm of deeptech, you better figure that out, or else the

---

87 "Introduction to Effective Altruism," Effective Altruism, June 22, 2016.

market will do so for you. When I say doing the "right thing," it is not about subjective desires but rather of quantitative and qualitative facts.

The simplest example of this is how it is estimated that the most efficient way to save a life is to donate anti-malarial nets to sub-Saharan Africa. Preventing malaria is your best bet because nearly 200 million people suffer from it annually, with up to nearly 800,000 dying. Donating anti-malarial nets is extremely cost-effective—it would cost around $3,500 to save a life from malaria using nets.[88] Or, it was found "if a typical American family were to transfer 1 percent of its income directly to an Indian rice farmer, it could double his happiness." If you want to increase universal happiness, why not do that as opposed to investing in productivity apps?[89]

That said, moonshots are large-scale. So, an appropriate example is the Green Revolution, which was begun by Nobel Prize-winning scientist Norman Borlaug in the 1940s. Instead of making incremental improvements in agriculture to feed a growing population, Borlaug invented high-yielding, disease-resistant wheat. It is said he has fed billions of people because of that innovation.[90] Think of that versus yet another food delivery app.

If you can execute within that context, and with a Stoic mind, then I believe you can make the moonshot within yourself, as you will be aiming for inner prosperity, blessedness, and ultimately eudaimonia. Always remember moonshots start with a wiser self. Once you master the self, you can master the world.

---

88 Derek Thompson, "The Greatest Good," The Atlantic, June 15, 2015.

89 Ibid.

90 Nobel Media, "Norman Borlaug — Facts."

THE PHOENIX WITHIN

"If things are not failing, you are
not innovating enough."

—ELON MUSK

A SLIGHT TASTE OF VICTORY

Most of us see the red flags. The natural disasters. The destruction of nature. The global warming.

Yet only a few of us find ourselves trying to chip away at the climate crisis. While the smaller-scale efforts are quite necessary, it's been a while since the last breakthrough that mitigated the acceleration of it. Project Foghorn was one of those.

Nonrenewable energy is a major component of climate change, but disrupting that is a rare sight. Nonetheless, in 2013, the X Moonshot Factory's Director of Mad Science and leader of the Rapid Evaluation team, Rich DeVaul, encountered a scientific paper explaining a way to extract CO_2 from ocean water through bipolar membrane electro-dialysis. In simple terms, there could be a way to make fuel out of seawater. Think about that—converting the ocean into energy. The uses for that are endless.[91]

He brought on Kathy Hannun to lead the effort once they realized the incredible impact this project could make. It directly fit the three-part blueprint for moonshot projects at X: a huge problem (climate crisis), a radical solution (using seawater as fuel), and a breakthrough technology (CO_2 extraction). Foghorn had the potential to affect every single industry and person out there for the better. It could be the biggest innovation in energy in decades.[92]

After validating their theories, the team forged forward to develop a "full-scale, cost-competitive carbon-neutral process for converting seawater into fuel."[93] Passions were high. People were excited. They were on the verge of something big.

Foghorn searched the entire globe for people working on the issue, from Idaho to Singapore. They consulted experts and research labs. They even held an international conference.[94] Yet, one day, their frenzied journey came to a halt. They faced four critical obstacles. The first two were obtaining industrial production amounts of carbon and hydrogen. The third was combining the two in an end-to-end, carbon-neutral system. The fourth and final challenge was achieving the first three steps in a manner that yielded fuel prices less than or equal to $4 per gasoline gallon equivalent within five years.[95]

---

91 Robert S. Huckman, Karim R. Lakhani, and Kyle R. Myers, X: The Foghorn Decision: Case 618-060, distributed by Harvard Business School.

92 Ibid.

93 Ibid.

94 Ibid.

95 Ibid.

In a quick pivot, they decided to work on hydrogen extraction because the carbon-neutral fuel production required it. It did not take too long for the project to identify a solid oxide electrolyzer cell that could extract H_2 from the ocean using electricity to divide the hydrogen and oxygen molecules in steam. It was yet another radical technology.[96]

Unfortunately, they did not have the resources to pursue it, so X decided to pair with desalination plants as a hydrogen source. That lead to another issue: there weren't enough plants that had the capacity to meet Foghorn's needs. Nonetheless, they doubled down and tried to do it themselves. Their spirits skyrocketed. They were so close to materializing a solution to the climate crisis.

But they never reached the moon.

RETHINKING FAILURE

DeVaul and Hannun ended up inventing a feasible technology, but it wasn't commercially viable. X leadership and the Foghorn team realized no matter how amazing the radical solution was, it could not compete with gasoline's cost-effectiveness. For reference, the seawater fuel would've been priced at nearly $15 per gallon. We all complain enough about gas, but imagine having to pay *that* much for it. In 2016, the decision to drop the Foghorn project was a tough one. The technology *worked*, but after years of persistence and determination, what killed it was the market itself.[97]

I recount Foghorn's story for an important lesson, however. It's that X viewed it as a success, no matter the results. They gave raises, vacations, and awards. They happily decided

---

96 Ibid.

97 Ibid.

to end the journey. It sounds odd for people to celebrate defeat, but that is the nature of scaling a startup. This is about rethinking failure, using it as motivation, and positively viewing it as a kill signal.

If you are serious about your innovation, you'd be lucky to succeed even 1 percent of the time. Thus, how you discover the other 99 percent and deal with it is the main factor. You cannot predetermine what path may lead you to the 1 percent success rate, but you can efficiently test and overcome the opposite 99 percent.[98] It's widely known the best founders are stubborn in the face of failure and are comfortable with uncertainty. Yet, as you sidestep the loss with this perspective shift, another thing becomes clear.

It's inevitable to encounter awe-inspiring ideas that turn out to be unfit for a sustainable business. Accordingly, you have to be open to killing projects once they pass the threshold of no viability. Yes, you heard me right. I may speak of "doing the impossible," but sometimes a business on the frontiers of science and technology is unattainable due to basic economics.

This is called being passionately dispassionate. It's the balance between knowing when to be enthusiastic enough about a venture and continue through uncertainty and fear while remaining detached enough to know when to stop the flow of effort, capital, and, most importantly, time.[99]

X's Obi Felten makes an analogy with the concept of inertia, or resistance to velocity. As Isaac Newton theorized, an object at rest stays at rest, and an object in motion stays

98 Teller, interview by Azeem Azhar.

99 Astro Teller, "Tips for Unleashing Radical Creativity," X, The Moonshot Factory (blog), Medium, February 12, 2020.

in motion, all unless acted upon. Similarly, once a founder is on their journey, it's easier to keep going than to pause. But the longer you keep going, the larger the opportunity cost. As passionate as one may be, they must have equal parts intellectual honesty and dispassion in the face of emotionally charged tenacity. They must accept when it's time to forego this moonshot for another idea.[100]

The science behind this is eye-opening. In the startup world, the idea passion is the biggest indicator of success is shoved down everyone's throats. Passion, passion, and more passion. We get it. But a 2009 study entitled "The Nature and Experience of Entrepreneurial Passion" by Melissa Cardon, et al., found sometimes, over-attachment to s leads to "obsessive, blind, or misdirected" patterns that interfere with effectiveness.[101] It seems intuitive, but not many people realize it. The fear of slowing down, the mindset of "destination or bust," can be harmful when dealing with the frontier tech.

CREATIVE DESTRUCTION

Another way to rethink failure is by recognizing how the best ideas are sometimes phoenixes that come from the ashes of a previously missed opportunity. Foghorn's findings led to Dandelion, a moonshot geothermal energy company that successfully graduated from X.[102] Founded and led by Kathy Hannun, the company adopted some of the team and the progress from Foghorn and pivoted toward a more commercially

---

100 Obi Felten, "How to Kill Good Things to Make Room for Truly Great Ones," X, The Moonshot Factory (blog), Medium, March 8, 2016.

101 Cardon et al., "The Nature and Experience of Entrepreneurial Passion," 551–531.

102 MacMillan, "Astro Teller, 'Captain of Moonshots' at Alphabet's X, Is on a Roll."

viable idea in which people can opt for geothermal heating and cooling as opposed to oil, natural gas, or propane. The switch from those fuels can even bring more than a 60 percent reduction in carbon emissions per home.[103]

The decline of one project marked the birth of another. It's poetic. Most importantly, it means failure isn't always a complete waste. Oftentimes, it may just mean the technology or science is not ready in that current moment in time. When the time *is* right, you can pick up where you left off, or you can even provide enough information for someone else to take charge. In Foghorn's case, Dandelion was so successful it officially spun out of X and became an independent business. It begs the question, how many moonshots have not seen the light of day despite the technology working? A quip in the deeptech realm says if someone these days comes across a novel idea, chances are some DARPA scientist or university researcher in the 1900s has already tried to pursue it. It just happened to get lost in the black hole of academia, or the market timing wasn't right. It's crazy to think how many inventions and innovations are out there already discovered but not brought into the real world.

The destruction of one idea leads to the creation of a stronger one. Everything that goes wrong in a startup's journey can lead to something even better. You just have to find value in the most unexpected places. You have to find the phoenix.

Nonetheless, every founder makes mistakes. It's inevitable, whether it is from not finding product-market fit, running out of funding, or from a failure to imagine failure, according to Josh Wolfe.

I specifically asked him what the distinction is between those who make the moonshot and those who don't. He said,

---

103 Dandelion, "Environmental Impact."

"The best [founders] are the ones that are intellectually honest because they don't fool themselves. They hire really smart people around them and are not afraid to have smarter people that inspire them. They hold themselves accountable—they're always quick with the bad news instead of constantly spinning positive news." Wolfe essentially says the same thing I've been hammering down: be positively negative, rationally skeptical, and heroically adaptable because success comes from the physical and intellectual growth rooted in that realization.

If you want a perfect example: humankind is the best moonshot out there. Statistically speaking, we most likely aren't supposed to be here. We survived for hundreds of thousands of years because of mutations, or the trial and error of DNA and natural selection, which allow us to test out new qualities and capabilities over time. Without these "mistakes," evolution would stagnate. We'd be unable to adapt to anything.[104] Errors are what genetically makes us the most biologically optimal versions of ourselves. Did we, as a species, falter when these "failures" happened? No. We progressed and became even better suited for progress.

So, maybe failure isn't always a bad thing—depends on how you view it.

104 "The Role of Error in Innovation," Farnam Street (blog), fs. blog, accessed August 29, 2020.

MOONSHOT
MENTAL MODELS

———

"Our use of the term 'moonshot' isn't literal; it's
more of an emotional blueprint. A moonshot is
about looking beyond where you can actually
see and envisioning an answer that doesn't
seem reasonable—and pursuing it anyway.
It's about doing things that sound undoable
but if done could redefine humanity."

—ASTRO TELLER

ON CLARITY

This section can go on infinitely because of its nature. In a
moonshot, company or otherwise, innumerable decisions
are made and countless new situations are encountered. To
navigate the complexity, the smartest minds out there rec-
ommend using mental models.

Mental Model (n): *"How we understand the world. Not only do they shape what we think and how we understand but they shape the connections and opportunities that we see. Mental models are how we simplify complexity, why we consider some things more relevant than others, and how we reason. A mental model is simply a representation of how something works. We cannot keep all of the details of the world in our brains, so we use models to simplify the complex into understandable and organizable chunks."*[105]

In simple terms, a mental model is a decision-making tool that allows one to apply proven logic and reasoning to any situation they encounter, no matter the relevance. They are a significant aspect of cutting through the entanglement between deeptech and worldly impact. Altogether, I call it "moonshot thinking."

In terms of startups, moonshot thinking is the core reasoning behind pursuing such a goal, much like how I describe radical creativity as the basis of the mindset. The former is about how entrepreneurs take the world in while the latter is about how they behave in it. This way of decision-making guides the genius and hero behind most moonshots as it delves into every field it touches: STEM, business, politics, social sciences, and the humanities.

Through my research and interviews, I found specific models shine above others. By no means is this is a definitive list, but it's a foundation. Thus, as an homage to the Apollo

---

105 "Mental Models: The Best Way to Make Intelligent Decisions," Farnum Street (blog), fs.blog, accessed October 11, 2020.

11 mission, here are eleven mental models for moonshot thinking. They are in this specific order to encapsulate key parts of the startup process. It starts with coming up with an idea, or first principles, and then processing it with risk-killing, anti-negativity, and advantageous divergence.

Next, leveraging luck, slingshot theory, Wriston's Law, and critical mass cater to how to position the business. Finally, compounding + Lollapalooza Effect, second-order thinking, and systems thinking entail a big-picture view after all is said and done. All are important, but you need the earlier ones to get to the later ones.

FIRST PRINCIPLES

First principles thinking is one of the more famous ones, as many of the best visionaries have admittedly used it. It's pretty ancient—Aristotle defines it as "the first basis from which a thing is known."[106] In other words, it's the systematic inquiry of a concept you break down into its elements and core assumptions. When you have a problem at hand, it means getting to its fundamental causes so you can reverse-engineer the solution from the ground up.

This way of reasoning is like a pyramid in which you need the foundation to get to the top. To do this, ask, "Why?" as many times as possible until you encounter the most basic piece of knowledge behind a certain idea. Why is it like that? Why did it react like that? Why was that the logic behind the decision?[107]

---

106 Aristotle, Metaphysics, quoted in: "First Principles: The Building Blocks of True Knowledge," Farnum Street (blog), fs.blog, accessed October 11, 2020.

107 "First Principles: The Building Blocks of True Knowledge," Farnum Street (blog), fs.blog, accessed October 11, 2020.

Thinking in first principles is about finding the root cause and meaning of anything. Elon Musk describes this as looking "at the fundamentals and [constructing] your reasoning from that, and then you see if you have a conclusion that works or doesn't work, and it may or may not be different from what people have done in the past."[108] Take battery cells, for example. Musk realized they were absurdly expensive. He asked, "Why?" It's because high-quality alloys are costly. But why? Well, because commodity prices determine them. But what's the material cost of them? Musk found it was around 2 percent of the market cost. Therein lies the solution—play around with the most basic components to make a cheaper battery:[109]

"...They would say, 'historically, it costs $600 per kilowatt-hour'... So, the first principles would be, what are the material constituents of the batteries? What is the spot market value of the material constituents? It's got cobalt, nickel, aluminum, carbon, and some polymers for separation, and a steel can. So, break that down on a material basis; if we bought that on a London Metal Exchange, what would each of these things cost? Oh, jeez, it's $80 per kilowatt-hour. So, clearly, you just need to think of clever ways to take those materials and combine them into the shape of a battery cell, and you can have batteries that are much, much cheaper than anyone realizes."[110]

---

108 Ibid.

109 Kevin Rose, "Foundation 20 // Elon Musk," Kevin Rose, September 7, 2012, YouTube video, 26:42.

110 Ibid.

The best companies start by identifying the first principles of a massive problem and solving it from the basics.

RISK-KILLING

This one is a wake-up call. I interviewed Contrary investor, Will Robbins, who said that "what sounds like a risk personally, probably isn't." Essentially, it's a check on if you are pursuing the right idea. Failing at a startup doesn't necessarily mean you yourself will fail. If you feel uncertain, that might signal a lack of inner conviction to continue it as setbacks arise. Any overwhelming hesitation is a red flag. Many deeptech founders have a relentless drive to build their solution despite everything working against them. They wholeheartedly believe in it. Thus, to them, pursuing the idea is not risky at all because the world needs it that much.

Obviously, it's human to sometimes worry about these things—don't be an emotionless robot. It reminds me of an idea by Josh Wolfe, who said "great founders and entrepreneurs are not risk-takers," but are rather "risk-killers" who "hate risk and seek to spot it and stamp it out."[111] Appropriately, one must eliminate the failures and the potential for them before they ever arise. This is achieved by taking a pessimistic approach to optimism, or actively shutting down any risk and dissatisfaction by excessive skepticism until one develops certitude with their approach. This "conditional optimism" is what propels the previously stated self-confidence that makes or breaks a successful journey.

---

111 Josh Wolfe, Josh Wolfe_volume2, 2017, distributed by Kevin Gao at Kinetic Energy Ventures.

ANTI-NEGATIVITY

The next mental model is also similar to optimism but in a different light. It's rather about avoiding negativity bias. In a *Wall Street Journal* entry, psychologist Roy F. Baumeister and journalist John Tierney depict negativity bias as "the universal tendency for bad events and emotions to affect us more strongly than positive ones." Everyone focuses on criticism instead of praise. We are ingrained in sensationalistic bad news and politics. We have a cognitive distortion toward cynicism because it's been a survival mechanism for most of human history. In terms of evolution, those who were cautious of their surroundings were the ones who survived. Mistakes were often fatal.[112]

This model exists because antagonistic views in execution, press, or anything influential are more striking than positive ones. That would translate into founders needing to focus on avoiding wrong rather than only aiming for right. "You get relatively little credit for doing more than you promised, but you pay a big price for falling short," says Baumeister. Similarly, his research into spousal interactions, customer reviews, and workforce behaviors found "it takes four good things to overcome one bad thing," a.k.a. the Rule of Four.[113] It's yet another tool for moonshot founders to think clearly when making decisions by minimizing the bad and accentuating the good. Capitalizing on this dual-sided model serves as a check on one's actions and motivations.

---

112 John Tierney and Roy F. Baumeister, "For the New Year, Say No to Negativity."

113 Ibid.

ADVANTAGEOUS DIVERGENCE

Just be crazy. Let me explain. Advantageous divergence is the culmination of contrarianism and audacity. Instead of developing these qualities, this mental model is more about operating productively and making decisions within an influx of ideas that are "advantageous because they are overlooked, discounted, misunderstood, too difficult to act on, psychologically uncomfortable, abstract, counter-intuitive, or have too long of a time frame to play out."[114] As a result, very few people know how to make the most of divergent notions.

The model becomes useful because it lacks widespread understanding, thus granting oneself an informational, analytical, or behavioral edge. Being unconventional takes courage because it always feels safe to go with the herd. That said, mainstream thinking will not lead to outsized returns. Outperformance and outlier results come from non-consensus ideas: "Extreme people get extreme results," says OpenAI CEO Sam Altman.[115] In the end, this model of reasoning can restore confidence in a moonshot founder as they deal with the ambiguity of trekking through the borderlands of science and technology.

LEVERAGING LUCK

Luck is, well, luck. Most people believe it is some external, uncontrollable force no one can harness. They believe it just comes and goes. I beg to differ. While that notion is true in some instances, it is not the full picture. A common pattern among successful people is maximizing serendipity and

---

114 Blas Moros, "Advantageous Divergence."

115 Sam Altman, "How to Be Successful," Sam Altman (blog), blog.
samaltman, January 24, 2019.

meeting opportunity with preparation. They let luck come to them.

Naval Ravikant, CEO of AngelList, says there are multiple types of luck: first is dumb luck; second is the notion "*Fortes Fortuna Juvat,*" or fortune favors the bold; third is the notion "chance favors the prepared mind." But the fourth type of luck is what Ravikant talks most about. It's the same luck moonshot founders take advantage of as a mental model. It's the idea "if you build your character in a certain way... then your character becomes your destiny," as Ravikant puts it. Making decisions in line with the person you are trying to be is key to making opportunities find you.[116]

In that light, you need to be gutsy when operating on the frontiers of deeptech. If you align your company in the best possible way (storytelling, culture, product, etc.) to make a genuine impact, the support, talent, capital, and customers will be drawn toward it.

You can't act normal and expect outrageous results. Make luck your destiny, not a far-off fortune. A positive take is the notion the universe is secretly conspiring to make you a success. Embodying that pronoia, not paranoia, is the basis of all luck-turned-destiny.[117]

SLINGSHOT THEORY

Gravity-assisted swing-by is the idea a spacecraft can use the gravitational pull and relative orbital movement of the Earth or any celestial body as a slingshot. If you watched *The Martian*, it's exactly what the Hermes rocket did to change

---

116 Naval Ravikant, "Make Luck Your Destiny," Naval (blog), nav.al, March 7, 2019.

117 Kevin Kelly, "68 Bits of Unsolicited Advice," The Technium (blog), kk, April 28, 2020.

their trajectory and accelerate back to Mars.[118] A spacecraft can use the downward pull of a body's gravity to whip itself around the body and change the course of its path. The key factor is the vehicle uses a larger entity's forces to sling itself into the optimal direction.

I was introduced to this model by Zenia Tata, Chief Impact Officer at XPRIZE. She related slingshot theory to what the best moonshots do: they grab onto the tailwinds of certain enormous trends and use them to accelerate their business. By "trends," I mean large-scale movements such as mass globalization, planetary health, or even human interconnectedness. They take advantage of the human nature surrounding these celestial-sized patterns and grow their company with it.

Even though such companies are typically—but not always—first movers, they don't act within the confines of an existing market. They create the markets instead by combining existing-yet-disparate demands and using them as a slingshot to reach millions, if not billions of people. However, doing so requires some sense of a massive force you can ride upon. For example, Boom Supersonic is making the first privately built commercial supersonic jet, but there has been an existing demand for it since the Concorde airliner days. Impossible Foods may have been a pioneer in the alternative foods space, but the underpinnings behind it were ripe for a moonshot, as many were disappointed with the treatment of animals and the deteriorating effects of the meat industry.

Just ask JFK and NASA. They did this perfectly during Apollo.

---

118 The Martian, directed by Ridley Scott (2015; Los Angeles, CA: 20th Century Fox, 2015), Cinema.

WRISTON'S LAW

Walter Wriston, former chairman and CEO of Citicorp (now Citibank), wrote in his 1992 book *The Twilight of Sovereignty* "capital, when freed to travel at the speed of light, 'will go where it is wanted, stay where it is well-treated.'" Wriston's Law can predict businesses' future performance and fortunes, as it asks, "Do companies (and countries) attract money and talent, or repel it?"[119]

Do heroic people want to work for you? Does the Invisible Hand, the force behind market equilibrium, bring talent and investment capital toward you? Answering these questions is an indicator of success and a mental model rooted in centuries of evidence. Take post-WWII America. New York City, Boston, and Silicon Valley became hubs of technology, science, and commerce because they attracted the right people who would thrive in the environment. It was even more evident in the 2010s, when the American job market attracted impeccable talent (unfortunately, we *might* be regressing now).

Wriston's Law is a testament to overcoming the monetary barriers of starting a moonshot company. It doesn't matter if you are poor. Bold things start with the mindset. But to make it an actual business, you need to treat people well: fair pay, quality of work life, benefits, health care, and much more.

CRITICAL MASS

Typically, the critical mass mental model is at the crossroads of chemistry and social dynamics. For the former, is it the least amount of fissile material needed to maintain a nuclear chain reaction. For the latter, it's the smallest sufficient number of "adopters of an innovation within a social system so

---

119 Rich Karlgaard, "Wriston's Law Still Holds."

that the rate of adoption becomes self-sustaining and creates further growth."[120]

Altogether, this presents a common misconception about deeptech. Doing cool things for the hell of it is great, but it doesn't make for a moonshot. They are still businesses that need to sell a product enough so they can sustain their large-scale efforts. Sure, we can invest into jetpacks, but are they really going to be well-adopted by the masses, as opposed to new therapeutics, which can benefit much of humanity?

Achieving that critical mass is integral for scaling a deeptech firm because relatively few people actually understand the inner workings of it. It really is the point at which you know you aren't just building for the sake of it, but rather for a tangible reason.

COMPOUNDING + LOLLAPALOOZA EFFECT

Albert Einstein once called compounding the eighth wonder of the world.

This cannot be any truer for moonshots, which operate on a pile of previous ideas and assumptions of a given field. Moonshots come from multiple disciplines that build off of each other to unlock new capabilities.[121]

This mathematical mental model is a way of getting value from a chain of previous events that snowball into something greater. It can even be interpreted as the basis of Kurzweil's Law of Accelerating Returns (example: Moore's Law). I like combining this model with the Lollapalooza Effect, which states "the parts in a system can interact

120 Gabriel Weinberg, "Mental Models I Find Repeatedly Useful," Gabriel Weinberg (blog), Medium, July 5, 2016.

121 "Compounding Knowledge," Farnam Street (blog), fs.blog, accessed October 11, 2020.

and influence each other, leading to shockingly powerful outcomes, where the whole is greater than the sum of the parts."[122] Altogether, huge things happen when the pieces of a system act in synergy over time. This is exemplified by William Brian Arthur's Law of Increasing Returns: instead of diminishing returns, companies can make market-dominating leaps by smaller competitive advantages that escalate in positive feedback loops. As participants try to outlast each other, they each become better. While only one wins (Facebook versus Myspace, Google versus Yahoo), almost all customers do so, too.[123]

In terms of deeptech, it's the decades of scientific and technological progress that come together to produce an even larger breakthrough—a moonshot.

SECOND-ORDER THINKING

This is all about wisdom as it relates to time: the present and near future. Second-order thinking entails being mindful of the consequences of your actions. While knowing what your decisions will result in is important, knowing what those results will, in turn, cause is equally essential. Everything has multiple order repercussions—think of it as the butterfly effect.[124]

Second-order thinking considers probabilities and payoffs, risks and rewards. To put it in perspective, think of a game of billiards or chess. Beginners tend to play move by move and strategize in the moment. Professionals think about

---

122 Blas Moros, "Lollapalooza Effects."

123 W. Brian Arthur, "Increasing Returns," W. Brian Arthur (blog), Santa Fe Institute, 2018.

124 Blas Moros, "Second-Order Thinking."

where their current decision will lead them, but also where they need to better set themselves up for the next move.[125]

This mental model imagines all possible, domino-affected repercussions of your decisions, which is not an easy thought experiment. My favorite example is Mark Zuckerberg. When he was making Facebook back in 2004, I'm sure he never expected his platform would be used by Russia to interfere with American elections. He probably never expected his product would contribute to the mental health crisis driven by social media. On the one hand, who could imagine those infinite-order consequences? On the other hand, Uncle Ben in the Spiderman series says it best: "With great power comes great responsibility."[126] Having that foresight and self-awareness is a must.

What will happen if ten million people adopt the product? Or what if there is a bad actor? Who is forgotten if your work scales quickly? What could cause people to lose trust in it? What may change culturally if people abuse it? What is the worst headline you can imagine? Many of these questions must be answered to engage in second-order thinking.

SYSTEMS THINKING

This one is vital because instead of wisdom in time, it's also about wisdom in space. Consider it game theory in action. It refers to viewing an entire ecosystem and each of the individual parts to improve a workflow, better understand relationships, or avoid unintended consequences.[127] The way I see this is about understanding complexity science as it

125 Ibid.

126 Stan Lee, Amazing Fantasy (1962) #15.

127 Blas Moros, "Systems Thinking."

relates to moonshots. It's a mere observation. I'm writing it as we enter a time of civil unrest while NASA and SpaceX are launching spacecraft. Here you have the hard work of incredible researchers and engineers making the impossible happen in space, but we can't even solve the problems we have on the ground.

Taking a systems approach to comprehending moonshot companies makes one thing clear: we must grasp the underlying factors at play. It's twofold. You must know exactly how every little change in a variable would affect an aspect of the business: for example, the effect of interest rates, R&D tax credits, or the cost of experimentation on revenues, cost of goods sold, wages, employee count, and so forth. Next, it's also about recognizing the impact of governments, academia, startups, venture capital, and the other players within the ecosystem. Especially when it comes to using science and technology for a global impact, there are so many things founders must be cognizant of in our rapidly developing society.

But it doesn't stop there. Where do we need moonshots, and in what form? How do we solve sociological problems such as racism or corrupt justice systems? What about technology-induced geopolitical turmoil? How can we encourage moral progress? How can we ensure ethicality when there are bad actors within the community?

Thus, in taking a systems approach to moonshots, we must expand the scope of what factors influence the purpose and building of one. We must consider societal impact, business, environment, labor market, local politics, history, infrastructure, justice, and the general entropy of every intermingling domain a startup touches. This is about ensuring that the future we are trying to build is one made with empathy, reason, sympathy, diversity, inclusion, maturity, fairness, equal opportunity, and basic human rights—with care, with good faith, with virtue.

The value of all of this? It's about systematizing moonshots. More importantly, it's about making *wise* companies. It's about empowering all of humanity in every way possible.

Like I said before, those are just some of the mental models emerging tech founders should employ. Use one or use them all; they will all lead to informed decisions as you execute. Maybe one suits you and another doesn't. But the value comes in combining many of them. If we apply the aforementioned mindset with these philosophies, we get a comprehensive profile on the intangible aspects of what it takes to make a moonshot company. These bold projects start with the mental and the emotional.

Now, we'll shift gears toward putting all of this into action. As we explore the tangible aspects, keep these themes in mind.

See you on the other side.

4

STRATEGY

SCHEMES OF COSMIC MAGNITUDE

THE FUTURE THAT NEVER WAS

One of history's most influential people was a futurist. In his essay "50 Years Hence," Winston Churchill described the state of scientific acceleration and then extrapolated it over a five-decade timeline. He was enamored by how humankind has progressed and regressed throughout history—how science causes a "new prodigious speed of man."[128]

> *The great mass of human beings, absorbed in the toils, cares and activities of life, are only dimly conscious of the pace at which mankind has begun to travel. We look back a hundred years and see that great changes have taken place. We look back fifty years and see that the speed is constantly quickening. This present century has witnessed an enormous revolution in material things, in scientific*

---

128 Winston Churchill, "Fifty Years Hence."

appliances, in political institutions, in manners and customs.[129]

Clearly, Churchill was a champion of optimism, heroism, and radical creativity. It's evident in his prediction of the world fifty years ahead of his time in 1932.

- *"New sources of power, vastly more important than any we yet know, will surely be discovered. Nuclear energy is incomparably greater than the molecular energy which we use today."*

- *"We shall escape the absurdity of growing a whole chicken in order to eat the breast or wing, by growing these parts separately under a suitable medium. Synthetic food will, of course, also be used in the future."*

- *"Startling developments lie already just beyond our fingertips in the breeding of human beings and the shaping of human nature."*

- *"Geography and climate will bow to the will of man."*[130]

The fifty-year prophecy got many things right—nuclear energy, synthetic foods, genetic engineering, terraforming, and so much more. It's eerily accurate. Churchill knew where we would go, and he was confident in it. But, one thing he could not do was predict what our reactions to these discoveries would be.[131]

---

129 Ibid.

130 Ibid.

131 Ibid.

*By observing all that Science has achieved in mod-
ern times, and the knowledge and power now in her
possession, we can predict with some assurance the
inventions and discoveries which will govern our
future. We can but guess, peering through a glass
darkly, what reactions these discoveries and their
applications will produce upon the habits, the outlook
and the spirit of men.*[132]

While this piece still rings true today, its context is a bit dis-
heartening. Churchill wrote all of that in 1931. He theorized
the world would achieve and scale those technologies within
fifty years—the 1980s.

In 2020, those are relatively new technologies.

It seems as though we've lost that collective hope. There's
evidence for how we're in an exponentially improving world,
yet it's also clear we are stagnated in deeptech advancement.
I believe we're somewhere in the middle. Only *some* fields
have seen that massive push. Most haven't.

Sure, we've had great leaps in information technology, communication, and digital. But what about energy?
Genomics? Nanotech? Construction tech? Health care? Since
the '70s, we got computers, the internet, and smartphones.
But we only recently began launching reusable rockets,
building fusion energy reactors, and putting in stronger
efforts against obesity.

Some were new discoveries, some were lost in academia,
some were underfunded—there are many excuses, but in the
end, we lack what we need most. Ironically, Churchill said it
best: we need to live in a world where the "schemes of cosmic

---

132 Ibid.

magnitude would become feasible."[133] Space metaphor? You know I appreciate a good one.

This is why I believe in moonshot companies. They're the startups that will put us back on the path we should've been on.

CRISIS MANAGEMENT

Now, we face crises just as important as putting ourselves back on track. The complex and polarized world is no stranger to extreme suffering: generational poverty, uncontrollable pandemics, inaccessible resources, crumbling infrastructure. Thus, moonshots have a dual mandate: fix deeptech stagnation and solve the world's most intractable problems. They do this at the sweet spot between being high-risk/idealistic and safe-bet/pragmatic.[134]

I spoke quite a bit about the mindset and philosophies, or the intangible aspects, of making moonshots, but now we have to put those into action. Only then will they be of any value.

This begs the question, how? How do we do that? It's not as easy as saying it.

Well, it's anything from finding a global problem to taking it in bite-sized steps to devising a crazy solution to solid execution. I'm not here to tell you how to start a company or run one. What I am here to do is inform you about the nonobvious strategies for getting started on one, so we can inspire a movement of impact-driven businesses.

With a combination of strategies from ancient times to Thomas Edison to the era of Big Science to now, I want to bring to light the visionaries within this crusade and how

---

133 Ibid.

134 X Moonshot Factory. Gimbal_v2.0_X (PDF). Accessed October 14, 2020.

they actually do it. How they act with the moonshot mindset. How they realize the noble end goal.

From geeks to polymaths, everyone has an essential role in invention and innovation.

We're out here producing heroes.

WE'RE GONNA NEED
A BIGGER BOAT

"Necessity is the mother of invention, and
I believe that we are at our most creative
when we embark on bold ventures."

—MIKE GRIFFIN, NASA ADMINISTRATOR

DIAMONDS NEED PRESSURE

We all know how dark a time World War II was: the rise of
fascism, the worry of world domination, the countless lives
lost. In the face of warfare, there was a group of incredibly
knowledgeable scientists and technologists who came together
to counter the Axis Powers.

Throughout history, there have been numerous
world-changing projects that resulted from wartime emer-
gencies. A prime example is the Manhattan Project. While
this was not a business, its roots point to a moonshot
idea-turned-reality.

The bombing at Pearl Harbor on December 7, 1941, dragged America into World War II. In response, President Franklin Delano Roosevelt commenced a military onslaught that heavily invested in defense technology and industry. He specifically gave $2 billion (about $35 billion in today's dollars) to the Advisory Committee on Uranium, which was renamed to the National Defense Research Committee, and then once again renamed to the Office of Scientific Research and Development (OSRD).[135]

As the Manhattan Project began due to a dire need to deter the Axis Powers, the OSRD brought together academics, theorists, physicists, engineers, and chemists to work on a weapon of mass destruction. Alongside this team was the Los Alamos Laboratory, which was already working on nuclear fission. The death toll increased exponentially. Morale was diminishing. The Allies needed a successful nuclear weapon.

The investment in science brought about that breakthrough.

July 16, 1945 saw the first atomic bomb's detonation at a testing site 120 miles south of Albuquerque, New Mexico. Then, August 6, 1945 saw the first bombing at Hiroshima, Japan.[136]

By definition, moonshots are beneficial to the masses. But, some of them have roots in trying times. The Manhattan Project is obviously a controversial example, but it shows how the boldest projects need sheer exigency. Extraordinary things come into fruition from hacking a societal pressure. Oftentimes, it's a situation so vexing it requires a large-scale effort. That's part of the attraction to deeptech, because alleviating climate change, feeding eight billion people, and

---

135 History.com Editors, "Manhattan Project."

136 Ibid.

keeping an aging population healthy are challenges worth dedicating a career to—and also markets that attract much attention from startups and investors.[137]

And, as the saying goes, diamonds are created under pressure.

FINDING THE CORE DILEMMA

It may be the lack of sustainable food sources, low household wages, exorbitant global debt, or the potential of an asteroid eradicating our planet if a nuclear war doesn't.

This includes even those who are the jurisdiction of governments. In fields where hyper-partisanship and special interests stymie proper regulation and progress, innovators are taking the reins and filling in the gaps where authority figures aren't.

You have deeptech startups bolstering national security through AI-powered everything, such as sensors, sentry towers, and drones. A plethora of moonshots are seeking to enact bold environmental change with cost-effective energy sources, computational agriculture, and computer-generated real estate development for smart cities. One of the largest areas in which entrepreneurs are doing the government's job is in health care. Whether it is therapeutics, vaccines, disease identification, or all-encompassing medical coverage, we have companies that can address anything. Fundamentally, they are derived from problems that affect millions, if not billions, of people.

Nonetheless, all of this transcends any one body's responsibilities—from space to the state of our oceans, from Alzheimer's to networks of underground delivery tunnels.

---

137 Portincaso, de la Tour, and Soussan, "The Dawn of the Deep Tech Ecosystem."

Finding the problem is simple. Think about what affects you, your loved ones, or your community the most. Mental health? Misinformation? Mudslides? In doing so, you must ask the right questions and frame them in first principles to tackle the root and not the symptoms. What sucks? How can you fix it? Is it good for me and the world? It takes some self-awareness and a lot of empathy. Moonshot founders genuinely fall in love with the issue.

To put this into perspective, take Tidal, a moonshot project in the works at the X Moonshot Factory. General Manager Neil Davé says the "biggest barrier to protecting the ocean—and our future—is that we don't know much about what's going on under the water." Tidal realized humanity is pushing the ocean past its breaking point and wants to protect it while sustainably feeding the nearly three billion people who rely on seafood. Other tangential troubles include pollution, hazardous fishing practices, and the acidification of water. Thus, they built machine perception technology that provides greater visibility into fish behavior with computer vision and environmental sensors that work effectively in harsh underwater conditions.[138]

Specifically, fish farming vastly contributes to the destruction of aquatic ecosystems. Tidal can now allow farmers to track their livestock's health and make smarter decisions about how to manage them, in turn reducing costs, waste, and the potential for disease. It's commercially viable, it's impactful, and it's radical. It solves the core issue, one most people overlook. And it's one that is easy to align with the mission.[139]

---

138 X Moonshot Factory, "Tidal."

139 Neil Davé, "Introducing Tidal," (blog).

As a whole, the underlying theme is finding the right problem to solve comes from deciphering the most fundamental matter one step at a time. Unfortunately, most people seek to solve the symptoms and are infatuated in doing so. A well-known example of this is how many want to "solve" climate change but fail to realize the crisis is a symptom of the Greenhouse Effect, which is accelerated by burning fossil fuels, which can be replaced by cleaner energy sources.

Deeptech founder, Naveen Jain, postulates another example about freshwater. The lack of freshwater is a critical issue to solve, so people are trying to find ways to make it abundant. On the surface, it seems to be scarce, yet we can take it a step further and ask, why is that so? One major factor is agriculture, which uses an exorbitant amount of it. Thus, if you satiate the enormous demand for water by farmers, you can cut down on its use. Then, Jain found within farming, most of the freshwater goes to raising cattle. So, what if we can nurture animals on our own terms—such as with lab-grown beef? There's a potential solution. With bio-factories and stem cell labs now able to create alternative meats, it is clear that one fix lies in synthetic biology, as Jain explains.[140]

This is all about finding secrets everyone is missing.

HACKING THE EXIGENCY

Considering the scale that such businesses operate at, it seems like the successful ones take advantage of the worldwide movements around the issues (remember the slingshot theory model). In other words, they hack the exigency around it.

---

140 Jain interviewed by Jim Kwik.

Take Kurion, a nuclear waste management company founded in 2008 by Gaetan Bonhomme, John Raymont, and Josh Wolfe. When everyone was chasing solar energy, ethanol, and biofuels, the team looked toward advancements in energy that no one was looking at—nuclear power. They considered the ever-increasing world population and economic growth, both of which alluded to an inevitable future demand for energy. The team thought about "what sources of energy would people realistically use in the future, how could we contribute to more available, less expensive options, and where is there an opportunity to build a profitable business?"[141]

Expert interviews, scientific research, and intelligence reports all pointed to harnessing nuclear energy. Yet, they realized a byproduct of nuclear operations, whether it's bomb-making or reactors, was the waste that ensued. What do we do with the waste? How can we clean it up? What if another Chernobyl happened? Wolfe found:

> "$1 of every $4 spent by the U.S. Department of Energy spent wasn't on 'green tech' but nuclear waste cleanup—$6 billion a year," and that "such cleanups could take 50 years or more to complete." At the time, it cost tens of billions of dollars to properly clean up and dispose of nuclear waste, and it was found that "hundreds of billions of dollars will be spent globally for decades to come" if nuclear waste management was not solved. With that in mind, Kurion invented a vitrification process to lower the lifecycle cost of

---

141 Lux Capital, "Kurion Overview," Lux Capital, January 31, 2016, YouTube video, 3:19.

nuclear waste management by isolating the radio-
active components of the remains and encasing it in
a stable formation of geological glass.[142]

I'm no energy expert, but that sounds pretty cool. Years later, Kurion had its big break—a total black swan event.

March 11, 2011. A 9.0 magnitude undersea earthquake shocked Japan and generated a massive tsunami that obliterated the Japanese shore. Homes, businesses, roads, and more were wiped out, and to make the situation worse, the power supply and cooling systems of the Fukushima Daiichi nuclear power plant were disabled. No one saw it coming, yet we all were horrified at the harsh reality. Kurion was one of a few companies selected to help contain the nuclear waste. In little time, they successfully designed, built, and delivered their technology to remove 68 million gallons of radioactive waste at a fraction of the previously perceived cost.[143] It was a win for science.

Just like them, stay wary for black swan events. Watch for the Sputnik moments, or realizations that people are falling behind in science and technology. Or, simply read the news. Use unfortunate events as encouragement and *make* the silver lining. Imagine if society as a whole could come together, put aside all differences, and ideate together—if not solve, at least support the efforts of one another.

In the end, we chase moonshots because they fix inefficiencies and ease dangers. So, when you are trying to find a grand challenge to solve, go bigger. So much suffering and death exists, yet so does much unmet commercial demand.

---

142 Ibid.

143 Peter Hebert and Josh Wolfe, "From Wasteland to Fund-Maker," (blog).

Most people think innovation is this asymptotic line that reaches some limit and will never go past it. While incremental innovators are needed, so too are the mavericks who push that limit and in turn uplift humanity. It's almost god-like.

At this moment in time, the world is trying to hack the exigency behind COVID-19. Vaccines are being fast-tracked. Politicians are touting fake cures. People are still dying. We need a moonshot.

That said, finding a problem is the easy part.

MOONSHOTS ARE FOR EVERYONE

"When something is important enough, you
do it even if the odds are not in your favor."

—ELON MUSK

IT'S NOT GAMBLING

There is one common argument against this book. Moonshots
are impractical!

They aren't easy, nor are they simple. But they
are *not* impractical.

If you look at history, it's easy to say most of these initiatives were a matter of "someone had to do it." Sure, someone
had to go to the moon. Someone had to revolutionize agriculture. Someone had to cure diseases. We have to stop waiting
for moonshots to happen. The future may be coming faster
than we think, but who are the people accelerating it? Sitting
around and admiring progress won't do anything. That's

why we have to overcome the mental aversion to making moonshots. They are actually not that impossible to start.

Remember how the Wright brothers were ridiculed for thinking they could fly, how JFK was laughed at for choosing to go to the moon? How no one thought college dropouts could build Apple or Microsoft? Many regarded these break-throughs as impossible and delusional before they were successful. Accordingly, many try to discourage founders from pouring their time into a project that might ultimately fail.

That's the nature of deeptech. You don't need that negativity.

Hitting obstacles is inevitable. I'm not going to sugarcoat it. I'm also not advocating for reckless optimism and fantasy. No number of motivational quotes or Gary Vaynerchuk videos can de-risk the possibility of a fiery crash landing. Maybe there are marketing concerns, regulatory matters, no product-market fit, dysfunctional co-founders. Maybe the problem is not yet solvable, or the technology hasn't reached the necessary heights. The fabric of deeptech is unrealistic. It's considered advanced and audacious for a reason. But there is a difference between gambling and moonshots. While the former entails pure and utter risk, the latter, if done right, is the opposite. It is rather a smarter, evidence-based, market-based, high-conviction bet.

No matter how implausible it seems, attempting such a startup means you're doing more for society than the majority of people. The odds may not be in your favor, but the upside to attempting it far outweighs any unfortunate downside. In doing so, an important consideration is if you're serious about it, you would be lucky to succeed even 1 percent of the time.[144] That's why there is experimentation, prototyping,

---

144 Teller, interview by Azeem Azhar.

and iteration. If you truly are passionate about solving some worldly crisis, a 1 percent success rate will not deter you.

Gambling entails taking risks for material gain. Moonshots entail putting a sound strategy behind societal impact. Gambling leaves its results up to chance. Moonshots make their own luck in a world of randomness and optionality.

Once you get over the mental hindrance of claims of impossibility, you'll realize how feasible it is to pursue an amazing organization.

NON-EXPERTS STAY WINNING

You read the subtitle. There are so many people I can mention. Most obvious is everyone's favorite tech bro, Elon Musk. He went from attaining bachelor's degrees in economics and physics to founding companies in fields such as internet software, financial technology, aerospace engineering, electric automobiles, artificial intelligence, neuroscience, and even tunneling. All of the companies are actual moonshots. Musk had no in-depth credentials in any of the fields he operates in, but he took his energy and ingenuity and channeled it into a fresh perspective that powered multiple bold initiatives.

That said, everyone knows him.

One lesser-known example is Laura Deming, a venture capitalist and modern-day polymath who focuses on life extension and the effects of aging. Growing up in New Zealand, Deming got interested in longevity at age eight. She was homeschooled and ended up teaching herself calculus, statistics, French history, and more. At age twelve, she moved to the US to work in a lab researching how to increase the life of Caenorhabditis elegans (C. elegans) by a factor of ten. She was successful in doing so. Then, Deming went to MIT to study physics at age fourteen, until she dropped out to start her venture capital firm, the Longevity Fund. Now, she has around $26 million to invest into high-potential life extension and anti-aging companies. She is fundamentally helping people live longer with synthetic biology, genetic engineering, math, and so much more. Her life's goal: biological immortality and finding the "lost Einsteins" of the world. I mention Deming because she was not a credentialed expert when she started. She had access to all the same resources most of us have, yet she went far with her skills and without a college degree.[145]

Another case is Naveen Jain. Among his many startups, one is Moon Express, which is building spacecraft to mine minerals on the Moon and eventually start colonies. He also founded Viome, a microbiome startup that is attempting to "make illness a choice" by finding balance within one's gastrointestinal system, which research links to being the birthplace for Alzheimer's, Parkinson's, autism, heart disease, cancer, diabetes, obesity, autoimmune diseases, allergies, asthma, and more. Ultimately, Jain was not a traditional expert in any of the fields at the founding of his companies—not space nor microbiology. Yet, his fresh perspectives challenged the norms of those industries, allowing him to incredibly disrupt them. Jain asserts experts are mostly fit for incremental innovations in their given fields because they "limit [their] own way of thinking" and do not question the foundational things that qualify them as an expert. That's definitely a hot take, but the point is clear.[146]

---

145 Henderson, "Meet the Teen Who Got Paid $100,000 to Drop Out of School."

146 Tom Bilyeu, "Naveen Jain on Why Curiosity Will Save the World | Impact Theory," Impact Theory, July 11, 2017, YouTube video, 50:25.

In response, he also says "when you come from outside the industry, you're able to question every single thing that people have taken it for granted." He portrays this by recounting how he went about creating Moon Express. Jain admitted he had no idea how to reach the moon when he decided to start this company. "When I started this space company, I had no idea how rockets work. I had no idea of astronomy. I had no idea of how orbits work," Jain contends. But he read voraciously and surrounded himself with the brightest minds in physics, aerospace engineering, and orbital dynamics and questioned them nonstop. Once he built a network and a minimal knowledge base in the field, he set his goal, and the rest is history.[147]

Experts and nonexperts are equally as necessary in scientific and technological progress. Yet, in arguing moonshots are practical, it's worth mentioning you don't need to be a credentialed master in a certain field to revolutionize it. There is a view that says nonexperts are better suited to take on moonshots. In short, some say the best founders are those who bring skeptical and diverse viewpoints. Traditional experts may consciously employ entrepreneurial myopia, or short-sightedness as it relates to startups: they know the generally accepted facts of a field and have to overcome a lot to go against them. They may engage in groupthink. Thus, "those who are down in the weeds are likely to miss the big picture," as Jain posits.[148] Essentially, being a credentialed expert in a single field may pigeonhole a founder into making incremental innovations as opposed to making huge changes. It's not an absolute, but rather a way to level the playing field. Anyone can be an expert, even if they are not a conventional one.

---

147 Ibid.

148 Ibid.

To add to this, one thing is that areas of study are becoming commoditized or obsolete relatively quicker. What constitutes deeptech has changed from the start of this sentence to now. The boundaries of what's possible are always moving forward. Therefore, anyone can gain actual mastery at the outermost edges because it is always evolving.

A glaring example of this is AI. A decade ago, artificial intelligence was something that struck fear into people's minds. The 2010s saw the widespread adoption of it, from manufacturing to drug discovery. The rate at which AI itself has improved is incredible. I know this to be true experientially, as my teenage friends are now toying with a next-gen AI called GPT-3, which is OpenAI's newest deep learning language prediction algorithm. Yes, it can now write eerily coherent *stories*. This technology was released in June 2020, and people younger than me are already creating products and companies with it. It's so new that most people have not even heard of it.

It's almost like bootstrapping—coming from "nothing," voraciously gaining knowledge, and taking advantage of unbiased perspectives. This is granted because of the "ubiquity of information" available on the internet and other accessible sources.[149] Together, it creates an ability for nonexperts to become well-versed in solving a problem without becoming ingrained and limited by its assumptions. A common pattern among the founders I researched and spoke to is they are all self-starters. No matter where they come from—whether poverty or wealth, college student or experienced professional—you do not have to be an expert

---

149 Naveen Jain, "Why Non-Experts Are Better at Disruptive Innovation," (blog).

to solve an extraordinary problem. You just have to start. If you're serious about it, you'll find your way.

Entrepreneurship is slowly becoming democratized.

STUMBLING ACROSS SUCCESS
Sometimes, luck does its thing. There are many inspiring stories of incredibly successful founders who began their startups after coming across an unmet need in the market. They find some proprietary, asymmetric insight that few, if any, have realized and take advantage of it. This effectively means *anyone* can be on the path to a moonshot.

To portray this, we can look to Brian Chesky and Joe Gebbia of Airbnb. Now, Airbnb does not fit the technical aspect of the three-part framework: after all, it's a landing page for living space rentals. But let's be real. The eighteen-billion-dollar company is one of the best-performing startups out there, whether it is emerging tech or not.

Back in 2007, the almost-broke duo moved to San Francisco and needed money to pay rent. When they heard a design conference was coming to the city, an idea popped into their heads. The hotels in San Francisco were nearly at maximum capacity, and many people were looking for a place to sleep during the event. The duo capitalized on this by bringing three air mattresses into their loft and listing them on Craigslist. Right away, three people booked the beds. Chesky and Gebbia realized they were onto something. Why were people so eager to live on an air mattress? It was because the town ran out of hotel rooms. Why did people seek hotel rooms? Because of the sense of trust they bring. You'd think that's a commonly known fact, but that proprietary insight was the start of Airbnb. They made living spaces more accessible through a trust system for private individuals to list and rent them, and in turn they spearheaded the sharing economy.[150]

Mike Maples, a venture capitalist at Floodgate, deduced the best startups come from "living in the future" and noticing something is missing and will be obvious to people one day.[151] That's why I admire Chesky and Gebbia's story. They did not have remarkably prodigious backgrounds or access to heavily guarded R&D. They were, in fact, against a hard rent deadline for their San Francisco apartment and desperately needed extra cash. It goes to show finding these inefficiencies can happen to anyone, and all that matters is if you take the initiative to pursue them or not.

IMPOSSIBILITY AND INEVITABILITY

Moonshots, however, are obviously not as easy as a typical company. If you take a step back, you'll realize the vast majority of people don't have the money to pursue one. Capital-intensive R&D and longer sales cycles mean there is a greater barrier to entry for frontier tech. Thus, affordability is an important consideration, whether it entails not having a risk appetite due to other priorities or not having the money to do so.

I came across a strategy that can relatively negate that worry. What if I told you that you could follow specific trends that point to something that's worth millions, if not billions?

These are called arrows of progress. They are trends at the crossroads of inevitability and the perception of impossibility. The way I interpret this is that these are patterns that follow societal progress as a whole. We all

---

150 Imelda Rabang, "The Airbnb Startup Story: An Odd Tale of Airbeds, Cereal and Ramen."

151 Farnam Street, "Mike Maples: Living in the Future," (blog).

love it when things get faster, more efficient, and cheaper. We all seek additional capabilities, increased productivity, greater knowledge, further control. Longer lives, healthier lifestyles, better communities.

One repeated example I always hear is Josh Wolfe's "Half-Life of Technology Intimacy." He cites how humans and computers are ever-so-slightly becoming one synchronized unit. Fifty years ago, we started with ENIAC computers that we would interact with through switches and plugs. Twenty-five years ago, we were introduced to personal computers that we interact with through a keyboard and mouse. After that, computers moved from desks to our laps as laptops. Then, phones went into our pockets, and AirPods went into our ears. Finally, we're in the age where contact lenses with displays are going into our eyes and where your thoughts and resulting intentions (neuron impulses) can noninvasively control external machines. Each step in this directional arrow was a moonshot in itself—a radically creative, culture-altering innovation that proved to be profitable. They also are not always some novel discovery, but rather the product of humanity's most fundamental needs projected onto the evolution of the product itself.[152]

What's also important is these step-changes are almost always considered impossible before they are made, and they are only considered inevitable by the masses after the fact. Thus, identifying those potential trends before they happen proves to be a gold mine of potential.

Evidence of this is how we tend to optimize the amount of energy density per unit of raw material, as we went from

152 Capital Camp, "Josh Wolfe - The Future, Now," Capital Camp, February 17, 2020, YouTube video, 51:45.

burning wood to using coal to splitting atoms to harnessing energy kites. Energy cannot be created or destroyed, but it *can* be cheaper, plentiful, and more accessible for the entire world. Or take the closing gap between physical and digital perception: from black and white television to high-definition screens to holograms, volumetric displays, and augmented/virtual reality. The trend seems to be that we want increasingly immersive experiences. When you recognize these arrows, it serves as a competitive advantage you can build moats around and protect at all costs.[153]

Moonshots are gradually becoming more practical. By no means are they straightforward, but there are ways in which one can demystify them and kill the uncertainty. The truth is people are not always in the ideal position to pursue such a risky venture. They may have families to take care of, or they may lack access to the right facilities. But the physical barriers to deeptech are slowly dropping, and I'd highly encourage people to move quickly in light of it. Once you get over the emotional hesitation, the world is at your fingertips—figuratively and literally.

---

153 Ibid.

MONEY TALKS

——

"Are you for real or full of shit?"

—ME

BAD BLOOD

I believe deeptech would be well-funded had it not been for one single company.

Some call it the worst scandal in Silicon Valley's history. Others call it the stroke of a mad genius. But one thing we have to admit is we all wanted to see Theranos win.

The story of this biotech startup is one right out of the playbook. In 2003, at age nineteen, Elizabeth Holmes dropped out of Stanford to pursue her dream for the future. She was young, scrappy, and charismatic like no other. She was the self-titled "female Steve Jobs."

Holmes founded Theranos, a portmanteau of "therapy" and "diagnosis," to revolutionize health care by creating a device that could perform nearly 250 tests, from cholesterol levels to complex genetic analysis, with just a single prick of blood. It was to be extremely fast and inexpensive—the be-all and end-all of diagnostic tools. It was a bold vision at the time, one definitely qualifying as a moonshot.

Right away, Holmes began fundraising vast amounts of capital for her R&D efforts. Some notable early investors are Henry Kissinger, Betsy DeVos, and Rupert Murdoch. She moved fast. The year 2004 saw a $6.9 million raise at a $30 million valuation. But, 2007 saw that valuation skyrocket to $197 million, and 2010 saw it rise to $1 billion. It reached unicorn status before the term "unicorn" was coined.[154]

This rapid blitzscaling was not without strict management. Holmes became notorious for creating a culture of paranoia and intimidation. She used threatening language. Her boyfriend became the company's president and also her figurative hitman. She was known to turn on employees in the blink of an eye. She even hired private investigators and allegedly paid big dollars to keep people quiet. The atmosphere she created demanded respect and loyalty at the expense of more than just being fired. Holmes kept every aspect of the company so secretive one whistleblower even said barriers were set up within the offices and labs so workers could not see the products they were physically working on.[155]

Theranos kept a low profile for the most part until 2013, when it announced it would go public. Then, the ball started rolling. Walgreens struck a multimillion-dollar deal to commercialize their diagnostic device. Next came a $9 billion valuation. High-profile media appearances. TIME Gala. A visit by Vice President Biden. A ground-breaking TED talk with the promise to make blood testing more affordable and accessible to all. Holmes had the utmost conviction and let

154 Connie Roff, "Everything You Need to Know about the Theranos Scandal."

155 Tim Ott, "Inside Elizabeth Holmes and the Downfall of Theranos."

her momentum take her to incredible heights. She was the poster child of entrepreneurship. The epitome of biotech. The one who would save us all.[156]

And it *all* came crashing down for one simple reason.

The technology did not work.

That's right. After all of that supposed progress, Theranos never actually had anything. There was quite a lot of pushback from the medical community throughout the company's rise, but Holmes was a master at manipulating people and the media. She could never lose. But it was all a lie. From the start, her startup never invented the godsend of a diagnostic device. Sure, it may have been too bold, but they fundamentally had fabricated the entire thing for more than a decade. Just think about it. They duped members of the public, investors, government officials, and the world as a whole.

In October 2015, the *Wall Street Journal* released a detrimental exposé on Theranos, effectively forcing Holmes to show she had nothing. It was a bombshell report, to say the least, and one Holmes vehemently denied. Nonetheless, she stepped down and was indicted on eleven counts of fraud and conspiracy. And with that, her company ceased to exist.[157]

Another thing that died was the general optimism about biotech and, in turn, deeptech. In my conversations with investors who were active around 2016, they noted a lull in movement within the domain as everyone slowly processed what had happened. In the end, a collective skepticism came from the fear of yet another Theranos.

---

156 Ludmila Leiva, "Keep Track of the Theranos Scandal with This Detailed Timeline."

157 Ibid.

ON STORYTELLING

I tell you the story of Theranos because although it was a complete fraud, they tricked almost everyone into believing it was real. The single reason for this was Holmes' charisma and storytelling abilities. That's all she had, anyway. She had no evidence her devices worked, so convincing people by word of mouth and social proof was the only solution.

Moonshots are as bold as Theranos was in theory. But that puts founders and the public in a challenging position. Will people believe in the startup? How do you know it won't be a fraud?

Most people overlook the value of narrative building. It allows entrepreneurs to impart their vision onto others while maintaining credibility and signaling trustworthiness. This is especially important for all deeptech founders due to the nature of their startup—something that is probably going to fail. No basis of comparison nor standard key performance indicators (KPIs) exist when going through uncharted territory. It's called impossible for a reason. As such, the typical investor would probably pass if they are not reeled into the vision and mission.

At its core, storytelling is the most timeless means of connection with another person. Brian Boyd, author of *On the Origin of Stories: Evolution, Cognition, and Fiction*, attributes stories to awareness: they are "a kind of cognitive play, a stimulus and training for a lively mind." They evoke such deep relations between people they lay the basis of perception, judgment, and reaction in conversations and interactions. Not to mention, they are elements of intelligence, cooperation, pattern-seeking, alliance-making, and the strengthening of ourselves as a species.[158]

---

158 Steve Denning, "The Science of Storytelling."

That is exactly what emerging tech operators must project when attempting to attract talent, money, and public support. You need to fire up hearts and minds. You need to evoke the moonshot mindset in the people you are speaking to as well.

Many believe private investing, finding coworkers, and garnering widespread support are deterministic, meritocratic functions. In contrast, the truth is they can come without having a business at all, nor a working product, nor an established team. On the flip side, well-established startups (a relative oxymoron) are constantly rejected by prospective supporters. The difference between a mere idea raising tens of millions of dollars in investment versus a revenue-generating startup getting passed on is typically how the founders communicate their narratives.

To put this into action, I want to bring to light deeptech investor Zack Schildhorn's three maxims of storytelling. First, you must make your audience care. This means appealing to their desires and painting a picture of how the outcome of your moonshot can impact them. For investors, it means making them "feel like they're going to make money by being a part of something special and sizable." If you present the breadth of the opportunity well enough, you do not need to see your pitch shrouded in growth and market statistics. The best stories emphasize "the scale of the opportunity" in such a magnetizing way "you can almost see the dollars at stake with hardly the dimension of the market size." To exemplify this, take the movie *The Game Changers* and how it tackled the incredible feat of persuading American men to switch to plant-based diets. There are countless pro-vegan documentaries and exposés of the meat industry, but these mediums are not massively successful. *The Game Changers* got millions of people to make the daunting switch by appealing to most

Americans' self-interests: athletics and the military.[159] It presents the scientific facts and dispels common misconceptions in the context of two things Americans love.[160]

Second, the story must be interesting. This is important because it stirs curiosity and lets the audience do the work: they "build ideas piece by piece" and "[put] those pieces together" in a way that "rewires" their brain. In essence, founders should spread the contagious enthusiasm for the future they seek to build. Take Auris Health, a robotic surgery startup acquired by Johnson and Johnson for $6 billion. Almost like a video game, Auris's flexible robots perform endoscopy by way of a handheld controller. As founder Fred Moll narrated his company's inevitable journey, Schildhorn recognized the potential upside of minimally invasive robotic surgery without Moll explicitly saying it. Curiosity is a powerful tool. Always remember that.[161]

Third, and most difficult, is you must make it human. This means finding a balance between professionalism and light-hearted authenticity and putting your best foot forward while portraying the battles you endured to get to where you are. Schildhorn relates entrepreneurs to superheroes because they "push the limits of reason and reality to make life-changing advancements real." But we as humans gravitate toward superheroes because they share relatability and lessons anyone can embody. Superman wouldn't be interesting without Kryptonite; Iron Man would be bland without his egotistical

159 "The Game Changers," directed by Louie Psihoyos (2019; USA: Netflix, 2019), Netflix.

160 Zack Schildhorn, "Lux Capital's Zack Schildhorn on the Power of Narrative for Entrepreneurs," Lux Capital, January 22, 2020, YouTube video, 8:36.

161 Ibid.

episodes. This is all about harmony and vulnerability, as that establishes empathy, humility, and more.[162]

To circle back, Brian Boyd articulates "the most successful storytellers apply themselves to the listeners' dilemmas—not just to amuse, but to make them fitter to triumph in the contests of life."[163] Don't over-tell (like Theranos), but don't be boring. This is the way to demarcate the mavericks versus the phonies. Ultimately, the narrative needs to be backed up by visceral creation.

There are no concrete rules in pitching, but rather patterns of success that are up for interpretation. It's not only about getting investments, but also attracting advisors, customers, and other key figures. It can influence audience perception by inducing product-market fit or building an enchanting brand image. To investors, it de-risks the heroic efforts by presenting a level of passion and credibility. And with sharing your vision and generating engagement comes the support of enablers such as experts, universities, and government figures. Having this village behind your moonshot, solely from your story, will do more than just fundraise. The synergy of all of those parties will forge a genius startup for years to come.

---

162 Ibid.

163 Steve Denning, "The Science of Storytelling."

THE BLEEDING-EDGE

"Any sufficiently advanced technology
is indistinguishable from magic."

—ARTHUR C. CLARK

CHILDLIKE CURIOSITIES

I grew up in an upper-middle-class part of New Jersey. It's
as normal as any small town gets. The shops, the restau-
rants, the little league baseball. Everyone was quite similar.
Most seemed to be living the nine-to-five life, with the vast
majority of the people I've met being in some sort of financial
services field.

That stood out to me because my dad was the opposite.
While I always wished he had a typical nine-to-five job, he
would come and go to his lab at varying times. I always
wondered what it was for, so he took me one day. I remember
each turn in the road and all the houses we passed. There was
a big dip in the tree-lined street, a stoplight that was always
flashing yellow, and a farm with the same couple of cows. It
was a peaceful drive, one that would calm you down.

Imagine an oddly long, beige building nestled at the base of a hill with two floors of forest-green reflective windows and a parking lot extending to the same length as it. As we arrived, I noticed a windy road with an "authorized personnel only" sign to the right. We drove past it. I could only imagine what it led to.

I still remember the first day walking into the building. The place smelled interesting—crisp, heavily air-conditioned, slightly artificial. As we walked down the hall, the fluorescent lights seemed to never end. I passed a huge world map, a conference room, and then I wandered right into my dad's place. It was an immense computer lab with vibrantly colored wires all over and hardware that made it look like a spaceship. I was a kid in a high-tech candy store.

To this day, my dad works at Bell Labs. The one every tech history nerd knows. The Bell Labs that made groundbreaking discoveries and housed Nobel Prize-winning scientists. I just didn't realize how awe-inspiring the place was until I matured.

Before returning home, we decided to go up that restricted road. It was kind of mysterious, not going to lie. But we came to an expansive field with a weird structure. It was Bell Labs' Horn Antenna. My dad was totally geeking out. He told me how some scientists found evidence of the Big Bang using that antenna and how his lab changed the face of astronomy in that very moment. I didn't know what he meant, but it was super cool.

After that experience, I thought my dad was a hero. But that wasn't all—he told me this one wasn't even the primary complex. He eventually took me to Bell Labs' main campus in Murray Hill, New Jersey. That visit sealed the deal. I fell in love with everything deeptech. One distinct memory I have was staring in the face of every bronze statue of the Nobel Prize winners they have in a little garden. I made it my goal to do something good for science and technology, like my dad is, and like Bell Labs continues to do—although they do it at a slower pace than they did in the last century.

Fast forward to high school, and funnily enough, I fell into the finance crowd. I sort of blocked off the childlike curiosity I had for deeptech for the sake of investment banking, Excel, and fancy suits. But my early appreciation for science came back once I got to college. There's something so galvanizing about deeptech because here you have people who are devoting their lives to shattering the limits of humanity's knowledge. It took me quite a while to realize that, but I'm glad I did.

THE NEW AND THE OLD

By now, you probably know what I mean when I say the "cutting-edge." It refers to the paradigm-shifting, transformative, fringe cases of hard science and tech. One significant pattern of deeptech companies is they often combine multiple domains of study. They use one to explain another. They are a patchwork of wisdom from one field that fills in a hole of another field. Here, you have everything from quantum computation and neuroscience, to materials science and biology for space colonies, to morphogenetics and AI for generative design.

On the one hand, you have the infinite capabilities technology and engineering can achieve. Yet on the other, you have the most foundational truths in our universe: biology, physics, chemistry, and all of their sub-branches. The cross-pollination of such areas leads to an abundance of synergy, perspectives, and possibilities. Suddenly, out of the complexity of multidisciplinary fields comes the research of the future. And, upon commercializing that come moonshots.

However, it's worth mentioning moonshots don't always have to be novel discoveries. Sometimes, they reinvent existing societal infrastructure. Technology itself acts as a "resource multiplier" that augments accessibility and capability. It might just be that what is most impactful is drastically improving legacy systems.[164] Think about the problems in public transportation. Construction. Buildings. Health care. Drug additions. Food. Cities. Communications. Companionship. Financial systems. Insurance systems. Legal and justice systems. Energy. ElderTech. Education. Governmental Services. Cybersecurity.

We need moonshots that will transform production and economics itself. Biotech-engineering investor, Arvind Gupta, calls this being on a "ruinous path of consumption." Gupta indicates if we add another two billion people to the population, "coal extraction and burning would spike from 7.32 billion tons per year to 12.3 billion tons by 2043." He remarks how "plastics production, at current growth rates, would skyrocket from 7.8 billion tons to 18.6 billion tons." Not to mention, "we pull a trillion fish out of the oceans every year—and have hit the upper limit—so aquaculture is going to have to at least double, from 106 million tons to more like 250 million tons by 2043." All of these future issues point to one thing: unchecked consumption with exhausting Earth's resources.[165] We need to step up as soon as possible to avoid such a reality.

Clearly, there's so much out there that is cracking under pressure. Fixing it may need new technical tools altogether,

---

164 Vinod Khosla, "Reinventing Societal Infrastructure with Technology," (blog).

165 Arvind Gupta, "The $100 Trillion Opportunity," (blog).

or just a solution no one's ever implemented before. Distinguished investor, Vinod Khosla, argues this reinvention and reformation is "chaotic, disruptive, [and] unpredictable with many failed attempts, but failure won't matter; the sparse successes will." He says "if there is a 90 percent chance of failure on a transformative project then we have a 10 percent chance of transforming the world."[166]

The things we need can come from fresh findings or the reimagination of what we have. They don't always reach unicorn status. Sometimes they don't even aim for an impact, yet the byproduct of their operations inherently does good in the world. For whatever it may be, deeptech is at its core.

LOW-HANGING FRUIT VERSUS THE ENDLESS FRONTIER

A long-standing debate in the STEM world has been if we're stagnated or not in terms of scientific and technological progress. One theory comes from economist Tyler Cowen, who suggests "economic development and technological innovation have reached a plateau, and unfortunately, we in America are only now just realizing it." In other words, all of the problems we face, from science to hyper-politics, point to the notion "we have been living off low-hanging fruit for at least three hundred years. We have built social and economic institutions on the expectation of a lot of low-hanging fruit, but that fruit is mostly gone."[167]

These "low-hanging fruits" are the preconditions that allowed past innovations and inventions to take place. They

---

166 Vinod Khosla, "Reinventing Societal Infrastructure with Technology," (blog).

167 Kelly Evans, "The Great Stagnation, Low-Hanging Fruit and America's 'Sputnik Moment'."

were the relatively plentiful environmental factors such as rapid growth and great prosperity. Now, we have exhausted all of the benefits: median household wages have stagnated despite prices skyrocketing, the rate of growth is flattening, and we are not meeting the issues our exponentially changing environment churns out.[168]

Yes, we are moving forward as a society, but by how much we are doing over time is the key. The 1900s saw an ecosystem conducive to breakthroughs, with the advent of globalization, better communication, and idea-sharing. Now, we've hit a point at which our general pool of knowledge is so utterly complex it's difficult to make something out of it. Therefore, we reached for the easy stuff instead of the harder things: much of our recent innovations are private goods that benefit a few individuals but not public goods that do the opposite—for example, productivity apps versus wildfire prevention, molecular elements, and drug discovery.

Putting this theory in the context of Vannevar Bush's *Science, the Endless Frontier* is interesting. The name is self-explanatory. Science is one area in which we will never reach the outermost edges.[169] Thus, if we have found and used all of the low-hanging fruits, what more can we do? How do we reach the furthest bounds, at which lie the answers to things we can't even fathom?

Cowen says we must "raise the social status of scientists."[170] Khosla believes we need to "[evangelize] market

168 Ibid.

169 Vannevar Bush, "Science, the Endless Frontier."

170 Kelly Evans, "The Great Stagnation, Low-Hanging Fruit and America's 'Sputnik Moment'."

participants" for visionary ideas.[171] Gupta explains that industries "mutate" when R&D is met with recognition and awareness.[172]

The pattern here is clear: progress will come when we make it desirable. Massive inventions and innovations will come when we are collectively inspired by it. And, moonshots will result when we invest into all of these forces and bring them together.

To conclude, I want to put this in perspective. Take the four-minute mile. Since the beginning of running, but more so at least since 1886, athletes worldwide attempted to break the four-minute mile. For centuries, it was considered an impossible feat. However, on May 6, 1954, Roger Bannister broke the four-minute barrier and essentially conquered the unconquerable in three minutes and fifty-nine seconds. The athletic world rejoiced, but the interesting outcome was that only forty-six days after the record was made, Australian runner John Landy broke it with a three-minute and fifty-eight-second mile. After that, athletes worldwide began breaking the four-minute mile until it simply became the norm. Once Bannister broke the collectively self-imposed limit, it proved to others it was, in fact, possible to do. Thus, a new mental model was born—one that shattered the literal and psychological limit. Two professors at the University of Pennsylvania's Wharton School, Yoram Wind and Colin Crook, found "the runners of the past had been held back by a mindset that said they could not surpass the four-minute mile. When that limit was broken, the others saw that they

171 Vinod Khosla, "Reinventing Societal Infrastructure with Technology," (blog).

172 Arvind Gupta, "The $100 Trillion Opportunity," (blog).

could do something they had previously thought impossible." There was no "sudden growth spurt in human evolution" or a "genetic engineering experience that created a new race of super runners." There was simply a mindset shift that exposed the "impossible" for not really being so far off.[173]

That's how I view putting the moonshot mindset into action, specifically at the connective tissue of the hard sciences, technology, and engineering. Once one person does it, many more follow. Likewise, once one innovator is held on a pedestal, more try to emulate them.

That's also how I view overcoming stagnation so both the traditional fields and the magical, cutting-edge industries exponentially advance. Just remember imperfect action will always trump perfect inaction. So, get started.

---

173 Bill Taylor, "What Breaking the 4-Minute Mile Taught Us about the Limits of Conventional Thinking."

SCI-FI MEETS SCI-FACT

"We don't read and write poetry because it's
cute. We read and write poetry because we are
members of the human race. And the human
race is filled with passion. And medicine, law,
business, engineering, these are noble pursuits
and necessary to sustain life. But poetry, beauty,
romance, love, these are what we stay alive for."

—JOHN KEATING,
PLAYED BY ROBIN WILLIAMS—
DEAD POETS SOCIETY (1989)

FICTION TURNED NONFICTION
October 21, 2015: the day the world met an unexpected trav-
eler. He dressed differently, talked differently, and definitely
did not fit in. There was something off about this kid. He
wasn't from the area, but he acted as if he knew everything.
He was amazed at the simplest of things: hoverboards, flying
cars, and the like. Those were everyday things. What was so
surprising about them?

Wearable tech, biometric scanning, personal drones, holograms, self-lacing shoes—*Back to the Future II* predicted it all.[174] It begs the question, how did the movie's writers know? How could they have possibly gotten such accurate depictions of the future? Think about it. Then, think about *Star Wars*, *Star Trek*, *The Terminator*, *Total Recall*, and even *Iron Man*. The writers of all of these movies were correct about some sort of future technology. They did not have any prophetic vision (at least, not to our knowledge), nor did they have the ability to time travel. It was pure ideation by extrapolating the present into the future.

It's challenging to pinpoint the origin of sci-fi. One place to start can be ancient religious texts, which tend to have qualities of it throughout them when considering what will happen in the future. Then we have philosophical texts that, in essence, embody what sci-fi is. For example, Plato's *The Republic* is an argument about what can possibly happen to society, justice, and humankind once governments become corrupt. In Plato's time, this text dealt with what could be and what may happen in the future.[175]

Sometimes, philosophical and religious texts use fantastical elements to portray key lessons and stories, but those elements can be considered sci-fi. German astronomer Johannes Kepler, famous for his laws of planetary motion, is said to have written one of the first works of "real" sci-fi in his novel *Somnium*, which presents a story of a daemon telling an Icelandic family about an island that, in reality, was the Moon. The text is one of the first works of art that seriously

---

174 "Back to the Future Part II," directed by Robert Zemeckis (1989; Universal City, CA: Amblin Entertainment Universal Pictures, 1989), DVD.

175 Helen Klus, "Imagining the Future: Why Society Needs Science Fiction," (blog).

considered science and lunar occurrences with the essence of reality.[176]

Since then, the genre went through scientific revolutions and philosophical enlightenments, and today it is the result of the ebb and flow of past society. It acts as a medium for taking in current realities and projecting them into the future. Sci-fi intertwines the sentiments of the time at which it was written, so there is much value in the futures artists envision, especially in the realm of entrepreneurship.

Hard science fiction embodies prediction, futurism, and practicality. Thus, there are so many parallels between that and moonshot companies. For whatever a person wants to imagine, their ideas have a place in sci-fi and emerging technology. And, they get to be as bold as possible—it is written to evoke wonder, thought, and possibility.

Moonshot founders are essentially sci-fi artists who build instead of writing or producing.

BRIDGING THE GAP

The gap between sci-fi (what is imagined) and sci-fact (what is manifested into reality) is shrinking. This may be because either our scientists are becoming more creative, or our sci-fi authors less.[177] Or it's evidence sci-fi is predictive in nature.

Many believe sci-fi is this cool-yet-unrealistic expression of fantasy, that it has no factual backing. That is the same case for moonshot companies. No one believes such an organization can tackle such grand problems until that company actually does. The perfect example of this is SpaceX. It had excessive opposition, as people believed commercial

---

176 Ibid.

177 CB Insights Editors, "Game Changing Startups 2019."

spaceflight is fantasy. They've clearly shown it is in fact real. But everyone knows that. Some lesser-known examples: remember the high-tech heads-up displays in *The Terminator* or even in the sci-fi video game and media franchise *Halo?*[178] Well, Google Glass emulates that as an augmented reality, hands-free device for manufacturing and other engineering domains. Another one can be Aira, an acquired startup that designs technology to grant blind or visually impaired individuals mobility and independence through live data streams from wearable technology, GPS, or the like. Here's one more example. The full-body-scan metal detector in *Total Recall* is what Evolv Technology devised—an AI-powered, frictionless threat detection platform that can scan 2.5 times more people than regular metal detectors with vastly more accuracy.[179,180]

Ultimately, sci-fi is often considered "too fanciful or not rigorous in thought," but if you look at the great deeptech advancements of the twentieth and twenty-first centuries, "they are often preceded by descriptions in works of science fiction written decades before," says MIT Media Lab researcher, Dan Novy, who also teaches a class entitled "Science Fiction to Science Fabrication." So, if this is true, why not leverage the creativity that sci-fi exudes and commercialize the advances that are inspired by it?[181]

---

178 "The Terminator," directed by James Cameron (1984; Los Angeles, CA: Orion Pictures, 1984), DVD.

179 "Total Recall," directed by Paul Verhoeven (1990; Culver City, CA: TriStar Pictures, 1990), DVD.

180 Capital Camp, "Josh Wolfe - The Future, Now," Capital Camp, February 17, 2020, YouTube video, 51:45.

181 Rebecca Rosen, "Why Today's Inventors Need to Read More Science Fiction."

I spoke with sci-fi writer, Alec Nevala-Lee, about this. We discussed the concept of bottlenecks and constraints. It's the idea that obstacles or limits on capabilities can be small, but they may have a disproportionate impact on some pathway or may be a source of inspiration as they force us to pivot for success.[182] Nevala-Lee spoke about how building the future is similar to creating sci-fi in that both are about exploring what may happen given current circumstances and how we get there—whether it is good or bad. Both must overcome constraints, but the best ones take those and adapt to them in a radically creative way.

Both find value in assuming that ideation and brainstorming about solving problems should be done 100 percent unbounded by reality: for moonshots, it may be commercialization hindrances, business plans, and stakeholder obligations. But, in terms of execution, the ones that interweave and overcome the bottlenecks regarding practicality while working backward from the preferred future are the ones that win.

Finding these parallels is super fun. Take the *Star Trek* tricorder and the Motorola Startac phone—the first mobile flip phone.[183] Another example is how Steve Jobs created the iPad, which is reminiscent of Captain Picard's PADD.[184] And FaceTime? It is eerily similar to the videoconferencing in *2001: A Space Odyssey*.[185] All instances had their fair share

---

182 Farnam Street, "Mental Models," (blog).

183 Star Trek: The Original Series, executive produced by Gene Roddenberry, aired September 8, 1966 – June 3, 1969, on NBC.

184 Star Trek: The Next Generation, executive produced by Gene Roddenberry and Rick Berman, aired September 28, 1987 – May 23, 1994, first-run syndication.

185 "2001: A Space Odyssey," directed by Stanley Kubrick (1968; Beverly Hills, CA: Metro-Goldwyn-Mayer, 1968), DVD.

of constraints, but they balanced imagination and realism well enough to envision a tangible entity.

My favorite example: superhero fans can see their favorite characters inspiring such startups. Similar to how Professor X uses Cerebro to find mutants in the world, there is a company doing something similar for drugs.[186] Variant Bio is a people-driven therapeutics startup that searches for genetic outliers that may uncover future therapies. They specifically look for individuals who hold extremely rare phenotypes to sequence and make drugs out of them. For example, they identified an isolated tribe with nine remaining families in South America who have a protein that increases their metabolic rate while they sleep. If processed and commercialized, this can help the seventy million obese individuals in the US by transforming the protein into a weight-loss drug, as people can eat a pill and have their metabolism burn fat at night. Variant Bio has also identified groups of people who are genetically immune to malaria, those who need only an hour of sleep, and those who don't feel pain. Like Professor X, they search for "mutants," but they rather do so ethically, with respect for diversity and the greater good for society—in true moonshot fashion.[187]

Some of the most far-out sci-fi and moonshot combinations come from *Star Wars*, a fan favorite. One is Auris Health, which created an advanced robotic surgery system directly inspired by a famous *Star Wars: Episode V* scene. When Luke Skywalker had his hand cut off, he had to undergo a robotic surgery to mend it.[188] The founders of Auris admitted

---

186 Stan Lee, The X-Men: Return of the Blob (1964) #7.

187 Capital Camp, "Josh Wolfe - The Future, Now," Capital Camp, February 17, 2020, YouTube video, 51:45.

188 "Star Wars: Episode V – The Empire Strikes Back," directed by George Lucas (1980; Los Angeles, CA: 20th Century Fox, 1980), DVD.

to channeling that scene into their work. Next, you have a company such as Looking Glass, which mastered volumetric display, or holograms: a means of communication used in countless works of sci-fi. The first thing it reminds me of is the Princess Lea holograms in Episode IV.[189] Finally, the Drone Racing League (DRL) is very similar to pod racing from Episode I, as both entities have man-powered vehicles racing in an organized league.[190] Essentially, fiction such as *Star Wars*, *The Jetsons*, and those from the Marvel and DC Comics universes are the best examples of how sci-fi positively relates moonshots—from ideation to execution.

SHARING FUNDAMENTALS

Science fiction gives life to deeptech. It also gets the creative juices flowing. That said, it's not always useful to chase the former. Sure, it's awesome to have jetpacks, but why invest in that when people still don't have clean drinking water? It's amazing to have houses on Mars, but why not houses for the hundreds of millions of homeless people worldwide?

I mention sci-fi in the strategy section for a reason. It's more than just a thought experiment about extrapolating horizons and speculating on the consequences of such courses of events. The point is that at the core of sci-fi and sci-fact, both must encourage the "ethical and thoughtful design of new technologies."[191]

---

189 "Star Wars: Episode IV – A New Hope," directed by George Lucas (1977; Los Angeles, CA: 20th Century Fox, 1977), DVD.

190 "Star Wars: Episode I – The Phantom Menace," directed by George Lucas (1999; Los Angeles, CA: 20th Century Fox, 2000), DVD.

191 Rebecca Rosen, "Why Today's Inventors Need to Read More Science Fiction."

Just as sci-fi does, moonshot founders and employees must explore the ethics and morality of the outcome they are pursuing. Especially when it comes to making the impossible possible in science and technology. Consider the defense industry. Military research has given us significant innovations and products (DARPA, ARPA-E, etc.). Relevant startups have also made leaps and bounds in the drone, security, and defense industries. But what happens when they are used for ethically questionable things? One example is Anduril, an AI defense company. They make products for border protection and intelligence, so immigration law enforcement uses their technology. Considering the controversies at the US-Mexico border, many conversations about ethics must take place.

There are even some instances of bad actors in science. Take the concept of eugenics, or the notion some races are biologically more superior than others. This was the foundation of the Nazi Party. It was also uncovered that key people in the history of DNA discoveries have evidently agreed with eugenics. It's crazy to think researchers so influential to science and humanity itself did so with horrible biases.

Another aspect of ethics is the polarization of progress. As we try to make more moonshots, we must remember there are billions of people who can't. It's not because they don't have the mindset or motivation. They physically and systemically are blocked from doing so, whether it is a lack of education or other basic necessities. Accordingly, we need to make sure the cutting-edge is not just for the haves but also for the have-nots. A moonshot is something that positively impacts millions, if not billions, anyway.

Sci-fi stories are also cautionary tales about the negative aspects of failing to "predict the human results that the technology will create." They allow one to experiment with repercussions and make changes before the deeptech becomes real. MIT Researcher Dan Novy explains "just as storytelling gives you more lives to live, speculative design or science fiction prototyping gives you more iterations to consider before your creation goes out into the wild and becomes hard to control."[192] In simple terms, we must stress the importance of preparing for the worst, considering the complexities of modern-day society and culture. Founders have to do the same thing.

William Gibson, author of the seminal work *Neuromancer,* coined the phrase, "The future is here—it's just not evenly distributed."[193] The future he speaks of is invented, in part, by moonshots all around the world. Ideas are everywhere—they just need to be vitalized. Sci-fi is a way of doing just that: inspiring people, predicting possibilities, and considering nth-order consequences.

We all can have a part in upholding an artistic license to make science fiction into science fact.

192 Ibid.

193 Goodreads, "William Gibson."

5

ECOSYSTEM

DIAMOND AGE

INTERMINGLING DECADENCE

The Athenian Golden Age, the Pax Romana, the Italian Renaissance—all marks of times in which art, science, politics, and philosophy blossomed. In which heightened humanism and critical thinking pushed forth development and knowledge. They were the products of not just one or two people, but rather an ecosystem of many.

These prosperous phases of past civilizations are considered moments of exponential advancement. If we bring those principles to the modern era, it may just be that we can induce a golden age of our own. Such revolutions in past societies occurred through the interplay of two core entities. One was the abundance of the self—art, philosophy, morals, ethics, health, and anything internally facing. The other was the network-driven improvement of complex systems—science, technology, government, education, infrastructure, and all things external.[194]

They are also defined by Type I and Type II Progress. The former comes from pushing the frontiers of capability

---

194 Anirudh Pai, "Golden Age & Polis."

with access to all the best resources and knowledge. In other words, expanding the future on the edges. The latter is rather progress without invention in times when the fruits of the labor uplift everyone. This entails distributing the future so everyone can reap the benefits from the bottom up. Generally, Type I is driven by developed nations and Type II by developing ones, although this is not an absolute correlation. Both go hand in hand. Neither is more important than the other.[195]

Together, golden ages occur when humanity flourishes with both Types in geoeconomics, social stability, and new cultures.[196] In contemporary times, it feels as if everything has gone haywire, almost like Murphy's Law: whatever can go wrong will go wrong. We seem to have slowed down the rate at which both sides interact. Music, art, and philosophy have not seen large-scale breakthroughs in a while. Morals and ethics are ignored in the tribalistic, power-hungry society we live in. We're in a mental health crisis. On the flip side, hyper-politics is screwing up America, as one extreme wants to regress, and the other wants change by destruction. Education has not evolved since past industrial ages. Infrastructure is unable to handle the inefficiencies of over-efficiency—we lack systemic resilience.[197] When was the last time we saw not linear but exponential improvement in anything?

While we romanticize the success of past golden ages, it's important to note they still had their problems: slavery, diseases, inequality, or war. But it gets me thinking: are we living in one right now? We may have undergone golden

195 Benjamin Reinhardt, "Type I and Type II Progress," (blog).

196 Anirudh, Kanisetti, "What Makes a Golden Age?"

197 Roger Martin, "The Virus Shows That Making Our Companies Efficient Also Made Our Country Weak."

ages of specific industries or art forms, but as a country, let alone the human race, will we ever reach one? It's hard to tell.

Sure, there are tiny sparks of progress as a whole, but nothing sustainable. It's been a while since the last major collaboration between both the self and the system. That said, I purposefully left out science and technology because I think that is our saving grace. They have the power to radically enhance society as a whole. Put two and two together, and my thesis is clear: moonshots can bring forth a golden age.

Instead of just a few, however, we need an ecosystem conducive to building these massive change-makers. This creates a cycle: support one moonshot company to make more of them. It's a collective effort, one that can fundamentally reverse the downward spiral we are heading in.

First, we need to spread the radical creativity mindset and push people to dream bigger. Next, we need the philosophies and the actionable steps to start the heroic journey. Now, we come to a point where we need all of the players in the environment to support the seemingly impossible missions of others.

When it comes to moonshot companies, those players are the startups, investors, government entities, and academics—not to mention the secondary nodes such as media, accelerators, and more. I'm going to tell you how this latticework of stakeholders uplifts deeptech and what we have to do to bolster it all.

It's a team effort. Our team is the country as a whole and every other if they choose to do so. Let's start our own renaissance, enlightenment, or even Diamond Age.

PEGASUS, NOT
UNICORN

———

"This is what will change the world...a ground swell
of people pouring their energy into manifesting
their 'preferred future' instead of being worn
down by disillusion and disappointment."

—MOLLY CARLILE

ADDED BONUS

I don't mean this in some sort of profound, metaphysical way:
your lifestyle is almost certainly influenced by space travel.

Since the Apollo missions, there's been a collective dis-
missal of space exploration as something that is "useless."
But in actuality, some major aspect of our lives is affected
by the push to go beyond our atmosphere. This is because
NASA realized the same technologies they invent to complete
their missions can also make them money on Earth: they're
called "spinoff technologies."

"Space exploration acts as a lens that sharply focuses the development of key technologies through the rigorous scientific demands that arise from pursuit of the near-impossible. When seeking to do things that have never been done before, especially in the unique and harsh environment of space, NASA creates new capabilities and makes new discoveries that may have been unlikely otherwise."[198]

It's safe to say NASA embodies the moonshot mindset (how fitting). The push to commercialize their cutting-edge R&D began in 1976 when they established the NASA Technology Transfer Program. Its impact is immense. Out of it came some of the most widely used products, such as memory foam, LED lights, artificial limbs, cochlear implants, and even power tools. Spinoff technologies are also used for LASIK and many other surgeries, firefighter and military tools, solar energy cells, architecture for supercomputers, and AI-powered personal trainers. They've also improved non-space fields such as medicine, transportation, construction, public safety, consumer goods, software, energy, and so much more that would easily add a couple of thousand pages to this book.[199]

While NASA is literally and figuratively shooting for the moon, your life is indefinitely influenced by their efforts.

The reason I mention all of this is for two reasons. One, spinoff technologies mark a solid strategy for making moonshots: break the lofty idea down into building blocks that

198 Douglas A. Comstock and Daniel Lockney, "NASA's Legacy of Technology Transfer and Prospects for Future Benefits."

199 NASA Spinoff 2020 (PDF), NASA, accessed October 16, 2020.

combine to add more value than the sum of its parts. Essentially, view "success" in smaller measurements while aligning them with the long-term goal. To exemplify this, we can look to Wing, a project from X that is developing delivery drones. At first, they sought to transport medical tools to remote locations, such as defibrillators for heart attack victims. But before they could do that, they had to build technology that could be trusted under dire circumstances. They took baby steps by aiming to deliver food first. At that point, it was not moonshot-esque. But they realized if Wing could master that while considering time, cost, and temperature, then they could easily transition to medical devices—the impactful end goal.[200]

The second reason is the commercialization of R&D is what makes a moonshot a company. It's great to make deeptech discoveries, but the utmost value comes from translating it out of academia or research labs and into the real-world market.

THE FUNDAMENTALS

After speaking with Seth Bannon, co-founder of the deeptech firm 50 Years, one thing became clear. Making the jump from lab to startup is tough, but the transition from entrepreneur to executive is more challenging. We're going to focus on the former first. It's all about going from great researcher to founder. At this stage, the work you do is relatively similar—figuring out a technical roadmap, rapid prototyping, and going from concept to product. This is where innovation happens.*

---

200 X Moonshot Factory, "Wing."

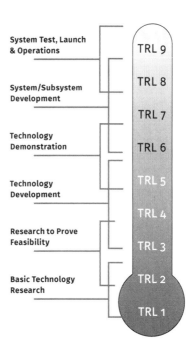

System Test, Launch & Operations — TRL 9

System/Subsystem Development — TRL 8, TRL 7

Technology Demonstration — TRL 6

Technology Development — TRL 5, TRL 4

Research to Prove Feasibility — TRL 3

Basic Technology Research — TRL 2, TRL 1

* Use NASA's Technological Readiness Level chart for reference.[201]

As for the latter, coming across something big in a university, private, corporate, or government lab is one thing, but surviving in the commercial marketplace is another. It's ruthless. You must become financially sustainable. Be socially aware. Be ultra-competitive. For scientists, this is where the principles of R&D clash with the business side.

Scaling from product to market comes with balancing science's doubts and entrepreneurship's conviction. Specifically, for academia, the culture is hierarchical, slow, conservative, and sequential. For startups, it's the opposite. Not to

---

201 NASA/Airspace Systems (AS), "NASA Technological Readiness Level Chart."

mention, they then have to hire a synergistic team. Create defensibility. Construct moats. Make sales. Market efficiently. Check metrics. Adopt regulations. Manage talent. Raise funding. Forge a moonshot culture. Find founder-market fit *and* product-market fit.

So long as the business model and impact are aligned, entrepreneurs become executives as they blitzscale.

I interpret the act of making these two transitions as navigating the literal and the philosophical. Moonshots walk on the fine line between an almost hedonistic individualism versus effective collectivism. This is about finding harmony between building cool things and solving bad things. It's more than getting your name in a journal. Now, you're shipping a product for the betterment of others. Rigetti Computing founder, Chad Rigetti, puts it clearly: "We're not doing this because quantum computing is interesting. We're doing this to cure cancer and to solve global warming."[202]

Moonshotting is not a PR stunt, as many businesses are guilty of. It's rather about contributing to the broader innovation ecosystem. But you have to love what you're doing and do it for yourself. Rigetti recommends using a key framework to maneuver these competing forces: are you pumping entropy in or out of your vision? Ideally, you want it out—the end goal is organizational clarity. Does management complicate things? Is the system too raw and complex? Or, is it conducive to growth? Is it crystal-clear enough for researchers to seamlessly become operators?[203]

202 Chad Rigetti, "Chad Rigetti at Startup School SV 2016," Y Combinator, September 29, 2016, YouTube video, 24:01.

203 Ibid.

Deeptech plays by different rules. It doesn't solely aim for unicorn status ($1 billion valuation), but rather profit and purpose: call it a Pegasus instead. They're here to create leverage. To capitalize on network effects and scale economics. To push the boundaries of what's possible.

ENTER THE DRAGON

Previously, I gave you strategies for actually coming up with a startup idea. You can do this from anywhere—from tech transfer offices to your garage—because what matters most is the destination: commercialization. As you can assume, it's a beast. Entrepreneurship is not for the fainthearted.

To slay this dragon, I found moonshot founders are risk-killers in two main categories: market and execution. These risks are universal to some extent, but mostly oriented toward frontier tech.

Market risk entails if anyone will want what you are building. It's everything you cannot control but can influence and outsmart. What's the competitive landscape? Why do people need your product over others? Will it sell? There are two perspectives on this. Some deeptech companies have minimal market risk because they are solving well-established problems in the public domain, like curing diseases or integrating robotic limbs into amputees. Others have a hard time because they might engineer solutions in commoditized industries with many players, although coming at it with a futuristic angle. Competitive risks also lead to scale risks, or how far out the company can reach. How fast can they grow? What is stopping them from achieving the demanded impact? Intuitively, these sorts of sciences and technologies take a longer time to de-risk with technical milestones that may not necessarily lead to more success. For example, successfully treating an ailment in a mouse does not mean it will work on humans. Creation is one thing; adoption and retention are others. Finally, legal risks may even affect the outlook and trajectory of the moonshot. Deeptech is typically more regulated than traditional companies. For example, biotech companies are entirely at the whim of the Food and Drug Administration in America and similar agencies elsewhere. They oversee the entire commercialization process, from idea to when they can start selling. Thus, determining the timing risk is also critical. Is this the right moment to launch? How long will it take to go to market?[204,205]

On the flip side, execution risks are everything founders can control. First, you have team risk. Do the operators work well together under pressure? What are the power dynamics? What is the allocation of equity and salary like? Do you optimize for experience, grittiness, or both? Next is product risk, such as if the technology will work and if it can be made into a sellable product. No one buys research papers or prototypes. They want actual, working goods. On top of that, intellectual property must be protected so no one can beat them at their own game. And, considering fundraising risks throughout the process is also highly recommended. While you can have an outstanding solution, the fate of the business rests on if you can sell your vision, whether for investments or customers.[206,207]

---

204 Celine Halioua, "Applying Tech Frameworks to Biotech: Key Differences," (blog).

205 James Currier, "The Hidden Patterns of Great Startup Ideas."

206 Celine Halioua, "Applying Tech Frameworks to Biotech: Key Differences," (blog).

207 James Currier, "The Hidden Patterns of Great Startup Ideas."

If founders can strategize against all of these risks, they are on the right path.

Like NASA and its spinoffs, the most impactful solutions come from the uncertain and unpredictable intersection of research and startup. They are symbiotic. In the end, making your way through this murky stage is a litmus test to see if the R&D is commercially viable and worth pursuing.

MONEY MAKES THE WORLD GO ROUND

"I don't want to be a driving machine on my daily
commute; I want to be driven by a machine."

—STEVE JURVETSON

SPARE CHANGE?
War-torn Iraq and Philadelphia, Pennsylvania, have more of
a link than most people think.

At first, Matt McGuire thought they were two completely
different worlds—he was a special operations soldier in the
US Army turned Wharton MBA student. While preparing
to become a tech investment banker, he came across a realization. McGuire observed some international students at his
campus would go to various pharmacies around Philadelphia
like clockwork. At first, he didn't think much of it.

Yet, as he saw more and more people do it, it just so happened they were all international students. He inquired and

found many of these people were buying pharmaceuticals from stores around campus to ship to their relatives overseas. McGuire found it was a regular thing. His peers would do so because their families did not trust the products in pharmacies in their hometowns. It turns out they were scared of counterfeit medicines, which are known to be ineffective or even to accelerate death. They did, however, trust Philadelphia medicines. Those worked. They needed them to survive.

That got Matt thinking. It reminded him of his time stationed in the Middle East. The "bad guy groups," as he called them, that he faced funded their terrorist operations in part by fake drugs. Thus, Matt delved deeper into this and was shocked at how pervasive the problem is. He found pharmaceutical companies lose about $200 billion in sales annually due to counterfeit products. But that's not all. He also discovered a million people die every year to them, with the vast majority of victims being women and children in developing nations.

"That's when Mainstream Matt died," he told me. He became obsessed with the crisis. It became emotional as well, as he found out his best friend's brother passed away in part due to fake chemotherapy. His life's mission became to fix this global problem. He began thinking of solutions and ended up teaming up with Texas A&M professor Dr. Johnathan Felts, who discovered a novel quantum property regarding waves, matter, and how light and energy react.

That was back in 2014. Since then, they've commercialized the nanotech solution. Matt became the founder of SafeStamp, a moonshot company that develops anti-counterfeiting indicators. The technology outperforms all conventional competing measures such as track-and-trace programs or holograms, both of which are easily faked by bad actors. It just so happens the only people who know about SafeStamp's nanotech in the world can fit in Matt's tiny San Francisco studio.

But he was not always based in Silicon Valley. Matt "lived like hell" in the early days: a dingy Nashville studio, showers at his gym, and working out of a coffee shop. At one point, he stayed up four days in a row to establish his startup and get funding. He was that serious about solving the oftentimes fatal problem.

I recount this story because initially, SafeStamp could not raise capital to fund the research and pay the team. From 2014 until 2020, Matt raised a minimal amount despite persistent efforts. When COVID-19 hit, SafeStamp's go-to-market plan was put on hold. And, they were negative in the bank account. They had everything for a successful early-stage startup, but they needed funding to pay the lab researchers and manufacturers. Without a capital infusion, SafeStamp simply could not progress. No sales or letters of intent could change that. But no investors understood the problem. None wanted to be the first investor.

That's what is crazy to me. Here you have a moonshot company with pre-sales, letters of intent, and working technology, yet no one was ready to take the investment risk. It had all the prospects any investor would want, but they all passed on it because they were not comfortable. An ironic instance was when one deeptech venture capitalist, who touts "saving the world" and funding the riskiest companies, did not put in the time to understand the product. For the longest time, the team was worried for the worst. It begs the question, how many moonshots go unfunded? How many world-changing businesses are there that ran out of runway? I dug into some stories and found incredible ones.

One that stood out is Escape Dynamics. Shut down in 2016 due to a lack of funding, the cutting-edge space startup achieved a breakthrough in space transportation: they constructed microwave propulsion technology in which a ground transmitter beams microwaves to a vehicle ascending into space. The microwaves heat a propellant such as helium or hydrogen to generate thrust. And yes, it's similar to the thing that is in your kitchen. It proved to be more efficient than the chemical propellants we use to this day.

But the team ended their push to revolutionize spaceflight, citing private investors were uncomfortable with funding it any more throughout the R&D phase.[208] This story is actually prevalent among hardtech teams.

Another story is that of Todd Rider, who developed a radical, broad-spectrum antiviral called DRACO. Viruses have ravaged humanity since the dawn of our species and killed billions of people since. DRACO was going to be huge: it's the closest thing to a panacea, a cure for all viruses. Starting his studies in 2011, Rider found DRACO could kill every virus in early tests: fifteen viruses in human cells and two in mice. Headlines went wild, with many calling it the biggest discovery in medicine since the invention of antibiotics. Rider was on top of the world.

But things went haywire because he was constantly rejected from labs, grants, and investors. He found himself in "the funding valley of death," as many wanted to see tangible efficacy in the medicine before giving money. But he couldn't show that without tests, which require funding. Ironically, Rider only needed a few million dollars to do the work—a minimal amount considering how software companies are

---

208 Jeff Foust, "Advanced Space Propulsion Startup Shuts Down."

given tens and hundreds of millions, sometimes billions, in funding. Not to mention, there was vast suspicion as to why pharmaceutical corporations were not supporting this, although they would obviously lose profits if DRACO was commercialized. It's surprising and confusing at the same time. Why isn't a cure for all viruses a priority?[209]

Now, I am not saying people should endlessly throw their money at moonshot companies that will probably fail. Nor will I ignore the fact most of these also need ten-year timeframes. I'm rather saying we have to collectively support endeavors like them. Maybe that means diversifying *many* investors' risk instead of burdening just a couple. Or, tapping into government reserves. In the end, we have to do whatever it takes to fund more moonshot companies.

As much as there is a moral imperative to make them, so too is there an importance to support them.

THE LANDSCAPE

There are two seemingly opposing fields of thought when it comes to funding moonshots. The first one is we need to finance them due to their capital-intensive nature and potential upside. I agree with that one, if you couldn't tell. The opposite is rather to fund incremental improvements and safer bets. As you can assume, the incentive structure of investing favors the latter. From corporations to angel investors to institutional venture capital, many of these collectives preach they solve the world's biggest problems by funding the best entrepreneurs. But do they really? That's questionable.

On a broader level, the mission-driven deeptech ecosystem is the moonshot ecosystem's foundation, as the latter is

---

209 Kevin Loria, "Huge Medical Breakthrough Can't Get Funding."

mostly the former companies but not always vice versa. These companies are R&D heavy and incur exorbitant costs before they have a sellable product. This makes the landscape as a whole notoriously difficult to tackle. It's like that for a reason. Frontier technology usually entails a longer "gestation" period, as product-market fit is challenging to secure.[210] Science problems aren't easy nor inexpensive to solve and have to navigate a ton of regulations. But the scene is changing—plus, software is getting pretty crowded.

These companies arise anywhere from garages to one of the more than 250 US research universities or more than forty federally funded R&D initiatives. And, they can come from any country around the world.[211] It all depends on how ambitious people are. That said, there's a spectrum between pure basic research, which lacks practical use, and pure applied research, or known science with only commercial purposes. The underlying concept is theory and practicality are not always mutually exclusive, and moonshots come from a balance between the two.[212] That's what investors look for.

GRAND SLAMS

Venture capital (VC) is the most efficient means of granting money to startups. The funny thing about it is while the industry is meant to invest in home runs (had to throw in a Wally Moon reference), they end up not doing so. It began by funding technically difficult, seemingly impossible businesses—in the 1950s, it was the advent of effective semiconductors

---

210 Different Funds, "DeepTech Investing Report (2020)."

211 Ibid.

212 Ibid.

and computers. Now, we have note-taking apps, luxuriously branded water, and robotic pizza-making services.

I'm not naive. The relatively "easy" stuff is where the most definite ROI is, although running a startup is not easy at all. But what about doing some good in the world? I'd say that's way better.

To do so, I found two things we can do from a funding standpoint.

First, we need to lower the cost of capital, or the required output startups must attain to grow. Simply put, this allows them to scale at faster rates, as there is less of a monetary hurdle for capital expenditures. There is a positive correlation between venture capital activity and the presence of deeptech companies, whether they are moonshots or not. For example, artificial intelligence/machine learning and life sciences are among the most invested sectors, with $4.6 billion and $19.3 billion in venture funding, respectively, in 2018.[213] It is no coincidence most moonshots come about from these two fields.

Therefore, we must increase the dollars flowing into these startups because free-flowing capital yields more R&D and much more audacious ventures. In 2019, there was about $136.5 billion invested by venture capital firms, with about one-fourth of it going to deeptech.[214] This number has to increase because more deeptech means more anomalies, or the ones that change the world.

"A low cost of capital is like a tractor beam for the future—it's like pulling twenty-year far-out projects into these twenty-month frenzied projects," says Josh Wolfe.[215] He's right.

---

213 The Engine, "2019 Touch Tech Landscape."

214 Ibid.

215 Wolfe, interview by James O'Shaughnessy.

Free-flowing money invents the future. And sure, it may be risky. Limited partners do not want to see their money go to waste, and venture capitalists have an obligation to them. It might sound controversial, but I would say forgo the investment into yet another social media company and focus on creating a positive impact. I dislike that it even is a "crazy" thing to say.

So how can we lower the cost of capital and in turn bolster the moonshot ecosystem? There are two paths. One is state-driven efforts. This entails venture capitalists and governments working together to fund moonshots. The second is narrative-driven efforts, or private individuals rallying people and painting a story that inspires people. Think about how Steve Jobs galvanized the masses. That means we need public agencies and private individuals to be entrepreneurial heroes by supporting and uplifting the startup scene.[216]

Second, VCs should engage in the moonshot ecosystem and nurture it, aside from solely looking for winners. Bolstering the environment can entail incentivizing, mentoring, or cross-pollinating the efforts behind them. That's the value of venture builders instead of purely venture investors. In the end, any more activity, besides increased funding, can have compounding effects on the entire landscape.

More activity equals more moonshots. That won't happen with solely focusing on yet another customer relationship management system or juice brand. Feeding eight billion people is a necessity. Creating and harnessing sustainable materials and resources is integral to human advancement. Keeping an aging population healthy is a must. Easing global warming, exploring space, building AGI, and designing

---

216 Ibid.

nanotechnologies—all are incredibly crucial to solving the issues that need more than just policy changes and social media awareness. Again, I get it. It's not easy. Also, CRMs and juices are important. But this is just a matter of priorities.

Similarly, VCs have to reinvigorate their efforts to prop up underfunded people. It's unfortunate seeing PhDs, doctors, and other researchers incessantly applying for elaborate grants they often don't even get. They are vying for relatively tiny prizes around $100,000 while we're seeing normal software companies raising tens of millions.

Women and people of color also fall into that category: recent studies found just 11.5 percent of all VC investments going to the former and even less to the latter.[217] The solution: let everyone participate in the economy. Empower everyone, open communication channels, and fix talent pipelines.

Bridging the capital gap and actively democratizing access to funds is the best way to achieve those priorities. It was found the coasts of America dominate the deeptech landscape, with the San Francisco Bay Area holding 31.6 percent of the deeptech market (percent of the total deeptech assets under management) and New York and Boston combined with about 32.6 percent.[218] If there were more investments into other entrepreneurial hubs, more would be incentivized to do audacious projects. A simple counterargument may be the startups that arise from the overlooked areas may not be sound investments (I personally disagree). But out of that disadvantage comes a potential opportunity: if the ecosystem is as collaborative as it can be, then even the most ignored areas can share the benefits. This comes from working

217 Kate Clark, "US VC Investment in Female Founders Hits All-Time High."

218 Different Funds, "DeepTech Investing Report (2020)."

hand-in-hand with research labs, corporations, government entities, universities, foundations, nonprofits, competitions, K-12 schools, media, and entertainment to distribute scarce capital in the optimal amounts.

VCs and those in charge of the flow of money have to bring more attention to moonshots. All of humanity would benefit from more investment dollars pumped into the ecosystem. A vigorous venture capital collaboration with the moonshot environment has unimaginable potential in unearthing those who will create the future.

THE OLD GUARD
IN NEW TIMES

———

"There are two giant entities at work in our country,
and they both have an amazing influence on our
daily lives... one has given us radar, sonar, stereo,
teletype, the transistor, hearing aids, artificial
larynxes, talking movies, and the telephone. The
other has given us the Civil War, the Spanish
American War, the First World War, the Second
World War, the Korean War, the Vietnam War,
double-digit inflation, double digit unemployment,
the Great Depression, the gasoline crisis, and the
Watergate fiasco. Guess which one is now trying
to tell the other one how to run its business?"

—ARTHUR P. BLOOM
(RESEARCHER WHO SAW A POSTER OF THIS QUOTE
AT VARIOUS BELL LABS SITES IN 1983 AS EMPLOYEES
PREPARED FOR THE DIVESTITURE OF BELL SYSTEM.)

TWO BIRDS WITH ONE STONE

You probably do not know one of the most influential figures in American technological history. We tend to think of Steve Jobs or Henry Ford upon first thought, but there was one unsung hero who ushered in a new wave of innovation in the 1930s and '40s. His name: Vannevar Bush.

As the globe marched toward an inevitable world war, it was clear the military technology of the past would be rendered obsolete compared to what the enemy had. Science had a vital role in turning the tide of the war toward the Allied Powers, but it wasn't so intuitive. It took a lot to get the government to realize that.

Enter Bush: a professor at MIT and inventor of the differential analyzer, one of the first operational computing mechanisms. Naturally, he was a huge proponent of deeptech as a means of winning the war. Even before Germany invaded Poland, he left academia to establish a national emphasis on research in Washington, DC. After the invasion, he approached President Franklin Delano Roosevelt about investing in science—the endless frontier that would hold the keys to a swift victory. Surprisingly, it worked. In 1940, the National Defense Research Committee was created to boost defense industries. About a year later, the Office of Scientific Research and Development was formed with Bush as its chairman.[219]

Bush was integral to the start of the era of "Big Science," the moniker used to describe the series of changes within the research community in response to World War II. The time was marked by large-scale, government-funded projects that pushed forward deeptech possibilities. In fact, Roosevelt

---

219 Michal Meyer, "The Rise and Fall of Vannevar Bush."

consulted Bush about what we had to do to take full advantage of these fields in supporting the military effort. In his famed letter, he wrote:

First: What can be done, consistent with military security, and with the prior approval of the military authorities, to make known to the world as soon as possible the contributions which have been made during our war effort to scientific knowledge? The diffusion of such knowledge should help us stimulate new enterprises, provide jobs for our returning servicemen and other workers, and make possible great strides for the improvement of the national well-being.

Second: With particular reference to the war of science against disease, what can be done now to organize a program for continuing in the future the work which has been done in medicine and related sciences? The fact that the annual deaths in this country from one or two diseases alone are far in excess of the total number of lives lost by us in battle during this war should make us conscious of the duty we owe future generations.

Third: What can the Government do now and in the future to aid research activities by public and private organizations? The proper roles of public and of private research, and their interrelation, should be carefully considered.

Fourth: Can an effective program be proposed for discovering and developing scientific talent in American youth so that the continuing future of scientific research in this country may be assured on a level comparable to what has been done during the war?[220]

Bush personally replied with a renowned essay that laid out the solutions. In summary:

Science can be effective in the national welfare only as a member of a team, whether the conditions be peace or war. But without scientific progress no amount of achievement in other directions can insure our health, prosperity, and security as a nation in the modern world.

The most important ways in which the Government can promote industrial research are to increase the flow of new scientific knowledge through support of basic research, and to aid in the development of scientific talent. In addition, the Government should provide suitable incentives to industry to conduct research, (a) by clarification of present uncertainties in the Internal Revenue Code in regard to the deductibility of research and development expenditures as current charges against net income, and (b) by strengthening the patent system so as to eliminate uncertainties which now bear heavily on small industries and so as to prevent abuses which reflect discredit upon a basically sound system. In addition,

---

220 Vannevar Bush, "Science: The Endless Frontier."

ways should be found to cause the benefits of basic research to reach industries which do not now utilize new scientific knowledge.[221]

A simplified answer: invest in science and take out the riskiness in pursuing R&D breakthroughs. Ensuring well-being, saving lives, reinvigorating research, developing talent—these aren't new ideas. We lacked sufficient means of winning the war, but more importantly, we did not have the necessary tools for human flourishment.

Vannevar Bush's story presents an important lesson on how government can bolster the moonshot ecosystem. Historically, it has always been the American government's purview to tackle new frontiers—the West, then Alaska, then outer space.[222] The same goes for technological progress. Over the years, they've set up incredible institutions such as NASA, DARPA, and ARPA-E to solve huge problems and make discoveries. They also established a plethora of heavily bureaucratic funding mechanisms that award the most innovative projects: Small Business Innovation Research (SBIR), National Science Foundation (NSF), and many other grants. Not to mention, there are individual departments and entire university systems that uphold their same goals.

There is no doubt our government has disproportionate power within the general entrepreneurial realm. On the one hand, it wields regulatory and legislative power. On the other, it has capital and access to any resource in the world. It's the ideal body for progress. It can incentivize

---

221 Ibid.

222 Ibid.

small businesses, promote economic development, motivate people, enable solutions, de-risk ventures, enact economic moats, ease regulatory hurdles, enforce competition, and so much more.

All of that is amazing. But why don't we see increased attention for emerging tech? Why keep throwing money all around instead of solving unjust incentive structures and power dynamics? Why is it we don't see much research commercialized? To make it worse, hyper-politics blocks us from changing the already flawed system.

In Ross Douthat's *Decadent Society*, it is observed "the peak of human accomplishment" and the "greatest single triumph of modern science and government and industry" was the Apollo 11 mission itself.[223] Yet since then, the relationship between government, academia, and startups has weakened. We haven't had an Apollo-esque win ever since. It's been more than fifty years.

We need that ingenuity back. We need to reignite that energy. It's pathetic that the original moonshot was our peak.

Changes must be made.

A FLAWED CULTURE

...in Order to form a more perfect Union, establish Justice, insure domestic Tranquility, provide for the common defence, promote the general Welfare, and secure the Blessings of Liberty to ourselves and our Posterity...[224]

---

223 Douthat, The Decadent Society: How We Became the Victims of Our Own Success, 10.

224 National Constitution Center, "Preamble."

According to the Preamble of the Constitution, to form a "more perfect" union, we must uphold justice, ensure tranquility, provide common protections, and promote the welfare of all. Doing so is pretty challenging. If that is true, and moonshot projects work on those already, then those in authority have an obligation to pursue such venturesome actions.

My thesis: if the proximate purpose of government is to address the Preamble's claims, then the ultimate purpose of government is to make moonshots. To do the hard things. To make prompt a golden age in which everyone can prevail.

As a society, we are not optimized for such companies. For one, it's because our military-industrial complex was linked to general innovation, Big Science, and moonshots in the past century. The Manhattan Project, post-WWII industry, advanced computing, Apollo missions. All were linked to military or geopolitical motives. Some were to flex our capabilities (Cold War), others were to prove them, unfortunately with much collateral damage (Hiroshima and Nagasaki). We needed lofty things to fuel the lies we told ourselves of American Exceptionalism. Sure, they had outstanding benefits in some instances, but it was rare to see purely good intentions. Obviously, I am writing this with a modern sense of morals and ethics, which evolve over time. There is bias, admittedly.

Nonetheless, my theory is sometime after the Apollo missions, there was a slight divergence between deeptech and defense. The link still exists, but the government sees non-military fields as less important: nearly $750 billion versus a mere $130 billion.[225,226] In fact, the American Associ-

---

225 Michael E. O'Hanlon, "Is Us Defense Spending Too High, Too Low, or Just Right."

226 Congressional Research Service, "Federal Research and Development (R&D) Funding: FY2020."

ation for the Advancement of Science found federal spending on R&D as a percentage of GDP dropped from more than 1.2 percent in 1976 to about 0.7 percent in 2018.[227] But, I'm not here to debate the military budget. What I am saying is in previous decades, its spending had spillover benefits to other areas, such as GPS, space rockets, and more. Now it does not. In this day and age, it does not entail solving the problems that moonshots do, for example creating intelligent medicines (Senti Bio) or nature-inspired, sustainable materials (Bolt Threads).

Even within military deeptech, we mostly focus on incremental technologies: faster fighter jets, larger explosives, or percentage increases in protection. Why not focus on fuel-efficient transportation? Smarter weapons? Bionics, drones, or AI to support or replace soldiers in danger?

Next, the government is structured in a way to favor short-term wins over long-term efforts. On a public office level, most of our leaders have narrow views because they solely seek reelection. They are not incentivized to support long-term entrepreneurial ventures. Representatives have elections every two years. Presidents every four years. Senators every six years. Moonshot companies engage in long-term thinking and strategies. Thus, there is a disconnect between what is a worthwhile investment and political goals. I am not advocating for longer term limits. We just need elected officials who care about the long-term as much as the short-term.

Third, the general governmental system does not support "anomalies." It is rather built for the masses. Moonshots follow unconventional paths. Therefore, those with power, influence, and money should be in a position to make them conventional.

---

227 Hourihan and Parkes, "Federal R&D Budget Trends: A Short Summary."

They should make entrepreneurial risk-taking less uncertain and use their reach to empower founders everywhere. It gets me thinking, how many people could have impacted the world but didn't because they were born into poverty or a lack of education?

This is a common complaint about SBIR funding specifically. People say while it's a great funding mechanism for startups, it is more attuned to small R&D shops focusing on agency-specific problems than to businesses with the potential for high growth. It favors top universities over relatively lower-ranked research universities. Most importantly, the SBIR program does not keep up with the speed startups operate at. The phases of the SBIR grant are time-intensive and bureaucratic, making it kill early companies with cycle time.[228] It's entrepreneurial Darwinism—a slow startup means a dead startup.

It is clear we are still incredibly ingrained in the military-industrial complex, and the emphasis and resources for private sector moonshots is declining. We need the opposite because everyone benefits from more financially stable deep-tech firms. The bottom line: we should adopt an entrepreneurial-industrial complex.

There is a glimmer of hope in that regard, however. I'll give two examples, one military-tech and one otherwise.

Take Anduril. It's is a new-age defense startup that leverages cutting-edge technologies to radically strengthen America and its allies' defense capabilities. Anduril is developing an AI-driven platform called Lattice-AI to spur forward advances in smarter border security, military protection, and critical infrastructure.

What I respect about Anduril is a quote they proudly display on their website: "We won't tell you that you're making

---

228 Different Funds, "DeepTech Investing Report (2020)."

the world a better place with ad optimization and emoji filters. We believe the most socially impactful thing we can do is help people in life-and-death situations make better decisions."[229] In that light, they reject what's sexy and lucrative and aim for making a change in the world—in moonshot fashion.

As I previously mentioned spillover breakthroughs, Anduril realized their solutions can be used for firefighters to tackle wildfires and ease the everyday concerns of regular people. If the government can fund and work with more Andurils, they can fulfill their part in the moonshot ecosystem and truly overcome its inefficiencies.

The other example is Farmwave, an AI-powered agriculture company that helps farmers with crop yields, harvest loss, plant count, pest and disease pressure, application coverage, weather, and other data-driven metrics. On its website, they present a USDA study that found Farmwave's technology would have increased the US corn revenue by roughly $1.4 billion.[230] Founder Craig Ganssle even told me Astro Teller himself said Farmwave is a necessary moonshot. Here we have private individuals stepping up to solve food scarcity and agriculture, which probably aligns with most countries worldwide. It's beneficial and profitable, yet we barely see any companies like it.

HOW TO IMPROVE

We've established our governing system isn't conducive to bold projects, but we also need actionable reforms.

From a monetary standpoint, everyone says we have to invest more into R&D. This also means better utilization

---

229 Careers, Anduril, accessed October 17, 2020.

230 Homepage, Farmwave, accessed October 17, 2020.

of our resources and collaboration across the ecosystem. It entails breaking down the barriers, such as easing unnecessary regulations in bodies and providing economic incentives for adopting smarter technologies. For example, tax credits for rooftop solar panels and electric car purchases. Progress comes by building an open and free society in which sophisticated financial and legal systems reward innovation.[231]

Then, we have to fix power structures. In essence, just allow entrepreneurs to actually be entrepreneurs, because moonshots are not led by big companies but rather by new entrants that change the rules of the game. Create governance-free zones that promote competition and ethical experimentation. Construct an environment that compliments human ambition. Or ensure a fluid job market. Also, do not pre-regulate unless there is a dire need. I'm not saying we need deregulation. Instead, we need more balance: make innovation a priority again.

Also, minimize the lobbying power of special and entrenched interests. It's widely known that oil and gas giants support politicians who stymie the shift to clean energy. Pharma corporations also pay massive amounts for leaders to protect their motives. One biotech founder who asked to remain anonymous said more than a decade ago, some pharmaceutical companies blocked a highly-effective cure for almost every disease out there. And yet, there were lobbyists and special interests who did not want that not to happen. Not to be a conspiracy theorist, but there are also instances in which scientists have "mysteriously" vanished or passed away as they were nearing some breakthrough. In a shocking

---

231 Steve Jurvetson, "Peter Thiel's Failed Futurism," Steve Jurvetson photostream (blog), Flickr, February 12, 2020.

revelation, one scientist told me she fears for her life in academia because of that very idea. In the end, we simply need an environment that is open to progress. Politics should bend to proven science, not the other way around. Otherwise, you get climate change deniers, flat-earthers, anti-vaxxers, and those who self-medicate with chloroquine. All of this can happen by electing politicians who understand the capabilities of deeptech.

People can't be future-optimists when they have to worry about their present state of well-being. This is where the government should step in because more moonshots will come when politics and social institutions improve. We need basic infrastructure, education, health care, and more. That would probably do more to stimulate deeptech progress than recklessly throwing money at it. Tech philosopher Anirudh Pai puts this a bit more eloquently. He found after the Apollo 11 mission, "the modern world has been dominated by sclerotic, parochial, and expensive institutions." The response? "Prioritize expansion over contraction, iteration over repetition, and construction over deconstruction."[232] Expansion, iteration, construction—it all points to making moonshots.

At the end of the day, the ideal governing structure is deeptech-friendly and wants to see *all* humans flourish. At times, these moonshots are even solving problems that arise from governmental actions worldwide. Bolster the ecosystem on the geopolitical side of things, and a golden age will come because these are the companies that truly make the world a better place...

...Sometimes, when governments can't.

---

232 Anirudh Pai, "Between Atoms and Bits," (blog).

AN ENGINE FOR LEARNING

—

"The hallmark of revolution is that the goals of the revolutionaries cannot be contained by the institutional structure of the society they live in. As a result, either the revolutionaries are put down, or some of those institutions are transmogrified, replaced, or simply destroyed."

—CLAY SHIRKY

CLASS IS STARTING

Almost every enlightenment throughout human history came from a rift in the consensus way of thinking. In classical Greece, philosophers such as Socrates, Plato, and Aristotle were the mavericks who advanced discourse around critical thinking, identity, logic, criticism, rhetoric, metaphysics, ethics, politics, sciences, justice, and so much more. Sure, they had their fair share of rejection, but it's safe to say they had a big part in sparking the revolution from a knowledge standpoint.

The ensuing prosperity of ancient Greece is reminiscent of the Renaissance—the rebirth of art, science, and social change that brought Europe out of the Middle Ages. It's the same story. Divergent thinkers had their time to shine and in turn caused a positive upheaval in most aspects of life. They made breakthroughs. They progressed intellectually. They produced masterpieces.

Golden ages occur in part by original thinkers who educate the masses. That's the key to an ecosystem conducive to radical improvements.

We all know education is the great equalizer, but its fatal flaw is it inherently is the antithesis of moonshots. As we try to teach all eight billion people, we do so in a mass-produced manner that is easier to spread but harder to personalize. A rising tide lifts all boats, but everyone around the world is not on one lake. There are rather mountains and valleys and rivers and dams that put some people in a better position to learn than others. Thus, you can't lift everyone with more education, but rather with one that appeals to their personal strengths and weaknesses.

I'm not saying everyone should be an entrepreneur. That would be counterintuitive. But we should design a system in which students are empowered to take action over their lives.

RIGHT THE WRONGS

Plain and simple, education is a human right. The first step is to make sure everyone has access to an outstanding one. Also, financial and nutritional literacy are an absolute must. But to make more moonshots, there are massive flaws in our country's system. While we are fortunate to even have one, it's skewed toward incrementalism. It's also the product of the industrial ages, a one-size-fits-all arrangement in which top performers are celebrated and low ones rejected. The internet made some significant enhancements to it, but only recently have we seen humankind truly accelerate education's efficacy and the resulting rate of socioeconomic progress.

The system as a whole is barely alright. We need it to be great. Thus, the next "internet" should be the perspective shift in which we inspire people to take on challenging missions. Even in the most ideal moonshot-esque combination of startups, investors, government, and general culture, the boost these give will only be minimal compared to if we can provide a means of educating generations to come. It's deeper than that, however. We need sweeping reforms because we need to train people on *how* to do the impossible. It all starts with enriching students' capacities and motivation to learn.

The biological basis of a moonshot education is neuroplasticity, or the physiological changes in the brain that happen as the result of our interactions with our environment. From birth to death, the brain is always reorganizing and bettering itself in response to evolving needs. The thing is these adaptations are hardwired for the rest of your life.[233] Thus, building the qualities of heroes, geniuses, and polymaths at an early age is a surefire way to produce the change-makers of tomorrow.

So what does this look like? First, it's about teaching students to be resilient, be curious, take initiative, fail fast and bounce back, and blossom their creativity in any form. Give them actually useful skills. Don't drill into them what works for another student. Or calculus. But really, rather

233 Naveen Jain, "Naveen Jain on Rethinking Education: Can We Use Neuroscience to Create Better Learners?"

improve the underlying foundation of entrepreneurial success. Next, make learning cool. Make helping others cool. Take away dumb distractions and busywork. Let students live. You know it's bad when the final bell to end the school day feels like an ecstasy high. It's sad to see an awe-inspiring concept made dull because of how it is shoved down one's throat and tested ineffectively on. Education should work for the students, not the other way around.

Training superhumans will transform every aspect of society because it implies better problem-solving, empathy, and diligence.

I got to speak with an educator who is trying to do exactly this. Navid Nathoo is the co-founder of The Knowledge Society (TKS), an accelerator program for students aged thirteen to seventeen. In simple terms, it's a school for future CEOs or even moonshot founders. Students who come out of TKS are building startups in industries from biotech to quantum computing to human longevity.

Nathoo explained that TKS's purpose is to create the right mindsets in students so they can unlock their full potential early on. He names six main factors behind his blueprint for "unicorn" students, as he calls them: mindset, wisdom, self-understanding, how the world works, fundamental skills, and exponential technologies. These six factors embody what it means to be a moonshot founder, even if you don't start a company. The most foundational aspect of this is the mind—not to toot my own horn, but he basically described my mindset chapter. Yet, in the context of education, he added he sees two camps of students: those who chase shiny objects and those who embrace fulfillment. This means some people are stuck on the desire for social acceptance while others move past that and work toward self-actualization. Thus, he teaches his students to be unfazed by others' perceptions and relinquish their dependency on social approval. In the end, teaching this to students at a young age can truly boost their chances of taking the risk of making a moonshot as opposed to anything else.

Nathoo also teaches his students about having a sense of self-understanding. Same story as before—his words validated my philosophy chapter. He said students must determine what impact they want to make and if it is the right one for them. Is it their most authentic passion? Nathoo adds to that and says, "If you don't know what makes you happy, if you don't know what you're good at, if you don't know what you're bad at, if you don't know why you're making certain decisions, then you're going to end up as a [cog in a machine], because that's the normal thing to do." In the end, Nathoo wants to nurture people who can impact a billion others, and he's made tangible progress toward that.

Everyone has their own personal passions that require the right environment and resources to pursue at an early age. And, while what makes the "best" mindsets and mental models may be subjective, I think it is safe to say everyone benefits from choosing resilience, bouncing back from failure, and thinking big. Applying a moonshot education to the deeptech ecosystem would be especially lucrative for humanity as a whole: the talent would have a domino effect on every aspect of the world.

An adaptive rising tide lifts all boats no matter where they start.

A BLACK HOLE

We're not done. Everything I mentioned is mostly geared toward schooling up to college. When you get to research universities, that's a whole different story. While the aforementioned critiques apply to this, the nature of academia makes it tough to materialize the potential moonshots they unearth. I already spoke about the difference between that and startups, but that isn't the main problem.

Academia itself does not favor the crazy and the bold. The individual researchers are amazing people, but as a system, everything is based on if you can get funding and publish papers. Thus, the "publish or perish" environment incentivizes playing it safe.

What's funny is unconventional research rarely gets funded despite statistically yielding better results.[234] But doing so breaks the academic culture built over decades. In fact, failures in science mostly go unpublished or undistributed.[235] Across many conversations, academics told that most labs don't publish failures and they end up running similar experiences in parallel because of this, wasting precious resources and time. Max Planck has a principle about this, being paraphrased as "science progresses one funeral at a time."[236] Essentially, progress occurs only when generations of scientists have different views. Within one generation, most follow the herd and accept whatever is generally seen as true. It's a slow environment, with centuries of practice ingrained into it.

A potential solution: establish a formal program in which prospective founders are partnered with researchers through tech transfer offices. Ideally, the scientists themselves would start the companies, but they can choose to stay and still see their breakthroughs become companies.

---

234 Okamura, "Interdisciplinarity Revisited: Evidence for Research Impact and Dynamism."

235 Karen Martin, "Publishing Failure in Science."

236 Jay Fitzgerald, "Does Science Advance One Funeral at a Time?"

Finally, academia has to become more open to dealing with heretical ideas. Thiel Capital managing director, Eric Weinstein, calls this the "distributed idea suppression complex," stating institutions block off ideas that oppose the norm, despite it being innovative and potentially impactful. You see it in politics, such as how left-leaning media ignored Andrew Yang in his 2020 presidential race and how right-leaning media silenced Ron Paul during his 2012 campaign. You see it in K-12 schools, where it's not acceptable to break from the norm of striving for straight As and a prestigious college. Within the research realm, this same pattern emerges.[237] Moonshots come from the opposite—the renegades of science and technology, trying to disrupt and radically improve systems with commercialized findings.

Ultimately, we have a long way to go with shifting education systems to a more moonshot-friendly aura. Nearly everything starts with this, yet many fail to see the sheer possibilities that come with it. Revolutionary people do not rise from carbon-copied establishments. Golden ages do not arise by thinking and acting the same way as always.

But simply make an environment that allows for founders to emerge, and I'm sure we will see vast success.

---

237 Eric Weinstein, "Eric Weinstein (Solo), Ep. #018 of The Portal - Slipping the DISC: State of The Portal/Chapter 2020," Eric Weinstein, February 13, 2020, YouTube video, 1:03:35.

CAPITALIZING ON COMPLEXITY

———

"Technological progress requires above all tolerance
toward the unfamiliar and the eccentric."

—JOEL MOKYR

WE'RE ALL FAMILY

We will never reach a utopia.

It's because the definition of it is always changing. What
is a utopia now will not be what it is ten seconds from now.
It's this end state we tell ourselves will come one day. It's
delusional.

That's why golden ages are vastly more realistic. Both
concepts are similar in that they depict an ecosystem driving
forth extraordinary progress and intellectual growth. These
eras are what fast-track us to where we envision ourselves in
an "ideal" world. The only reason for that is the interaction
between all parts of society that somehow ends up producing

synergy unlike ever before. That's the magical solution to making moonshots. The dynamism, the combinatorial creativity, compounding knowledge. That's what we need so badly. It should not be siloed to the well-educated. It should be democratized. It *needs* to be.

There is a relevant benchmark for this. It's called the Kardashev Scale—a method of measuring a civilization's level of scientific and technological advancement based on the amount of energy they are able to use. Soviet astronomer, Nikolai Kardashev, theorized there are three types of realms.

- Type I: Planetary civilizations that can harness all of the natural energy circulating throughout it. Attaining this would entail a full shift to renewable or fusion energy and full control over nature itself.

- Type II: Stellar civilizations that can harness the power of the civilization's planet's solar system, including its sun. Achieving this would need Dyson spheres, megascale engineering, and interplanetary travel. To put it into perspective, it's probably what ants think of humans.

- Type III: Galactic civilizations that can harness the energy of an entire host galaxy and every celestial body in it. It's hard to describe this, but these species would be able to use black holes, gamma-ray bursts, and so forth.[238]

Other theorists have gone on to say there are Type IV civilizations, or those that can harness the universe itself and

---

238 WatchMojo, "What If Humanity Was a Type II Civilization? | Unveiled," Unveiled, July 22, 2020, YouTube video, 8:01.

even parallel ones in a god-like manner. It's some crazy stuff. I mention this because we're sitting around a 0.7. After millions of years of humankind, we still have not reached Type I. It's no easy task, but it serves as a tangible goal we can meet within the next few centuries.[239]

That is, if we start making more moonshots.

See, you're not alone in taking on a noble pursuit. Deep-tech wonders don't happen in vacuums, but rather with a collection of parties helping founders cut through the noise and complexity. I call it a "moonshot ecosystem." The core infrastructure of this is the relationships between each player. If we can genuinely strengthen each node, we can create greater supercompanies. It's the butterfly effect, but only that the entire world exponentially improves with problems solved and humans thriving.

To better understand this, I break all of it into primary and secondary participants across three contexts. Primary participants include those from the aforementioned sections: startups and corporate lab projects, investors, government entities, and academia. Now, I want to highlight the second-ary participants, or incentivizers, accelerators, nonprofits and foundations, and the media. As for the contexts, there are geographic, mission-driven, and sector-related commu-nities within the broader ecosystem. This essentially means sub-ecosystems exist that are centralized by location, a col-lective goal, or an industry.

These environments rely on network effects (on steroids). They become incredibly more valuable with every added or strengthened party to the symbiosis of talent, expertise, ambi-tions, policies, and other maneuvers. You can even consider

---

239 Ibid.

it as a culture, a pseudo-religion. Think about it: if one party employs a forward-thinking mentality, the energy will bleed into others that it works with. It's contagious.

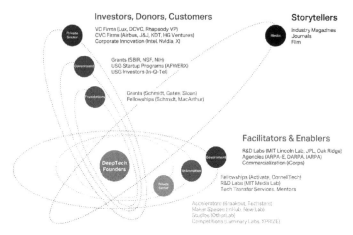

The Deeptech Ecosystem by Different Funds[240]

CONTEXTS

Geographic contexts are the most literal translations of what it means to have an ecosystem. Silicon Valley (SV) is a prime example of how the interconnectedness of startups, universities, research institutions, and entrepreneurship-friendly governments yields an innovative, extremely lucrative area. It boils down to three main factors. For one, SV has Stanford University, University of California, Berkeley, and a plethora of other outstanding universities and affiliated research organizations. This provides talent, experimentation, R&D, and more while training a new batch of students year after

---

240 Different Funds, "DeepTech Investing Report (2020)."

year. For two, SV has a relatively startup-friendly government that oversees small businesses, strategically implements fair regulations, and de-risks the process of starting companies. In 2020, this is widely debated and possibly regressing, but it historically is true, nonetheless. Three, SV is a tight-knit community with vibrant arts and entertainment scenes. Hubs need to use their soft power to create a magnetizing culture that not only brings in but consistently maintains attention and ideas. Sure, you need the academia and professional institutions, but the informal environment brings life while the former provides the structure. Altogether, the harmony of all these nodes socially engineers a geographic area ripe for bold ventures.[241]

Imagine if this was the norm worldwide.

Next, we have mission-driven ecosystems, or those that take a multidisciplinary approach toward a vast collective goal. One of the biggest ones is that of the fight for environmental protection. Companies such as Commonwealth Fusion Systems are taking the energy angles; those like Tesla are taking the automobile angle. Government policies are a major contributor to the fight against climate change—although maybe not nationally—such as the Paris Agreement. Venture capital firms such as Breakthrough Energy can support startups that fight for environmental longevity. Nonprofits such as The Ocean Cleanup are doing their part in cleaning the oceans. And there are a multitude of research institutions that are fortifying our knowledge in this space.

Finally, we have sector-related ecosystems, or those who operate within a certain industry but do not necessarily have the same goals. The key distinction is it's a collection of

---

241 Scott D. Anthony, "Entrepreneurial Hubs."

nodes that push for a specific field of study as opposed to key players from various areas coming together. For example, think about the healthcare sector. Incredible players within it seek moonshots in different aspects of it: Recursion Pharmaceuticals is accelerating drug discovery, Kallyope is seeking to tackle the gut-brain axis, and Verily is easing addictions, mapping the immune system, and more. On top of that, the entire government-backed insurance industry contributes to the healthcare sector alongside numerous research institutions. From collaboration to support, funding, and scaling, these communities are only as strong as the parties who populate them.

SECONDARY PARTICIPANTS: THE ENABLERS AND FACILITATORS

First off, we have the incentivizers, or the organizations that help people pursue ambitious targets.[242] The go-to examples of this are accelerators, incubators, and venture studios. There are competitions such as XPRIZE, which encourage moonshots in light of a certain problem for a sizable cash prize. Not to mention Hello Tomorrow, Singularity University, and more organizations that spread awareness. Altogether, these enablers de-risk the process by granting money, networks, and shared expertise.

Similarly, we have nonprofits and philanthropic foundations such as the Bill and Melina Gates Foundation, Schmidt Futures, and the Chan-Zuckerberg Initiative. They are additional lifelines for moonshot companies by means of funding, mentorship, and scaling. But they also are not necessarily driven by profits. They typically have ample funds on their own, so they're more geared toward positive impact.

---

242 Different Funds, "DeepTech Investing Report (2020)."

Finally, we get to the media. It's the bridge between bold objectives and public perception, just like how JFK's moonshot speech was a PR stunt to garner support for the expensive Apollo missions. Motives aside, news, social media, and entertainment do the same thing. They can inform and find prospective users or can even evangelize technology to inspire the masses—think of the effect *The Martian, The Social Network,* and *Interstellar* had on movie-goers.

IT TAKES A VILLAGE TO RAISE A MOONSHOT

We can be so much better. The things we have in place are pretty adequate. When I say, "We have to bolster the ecosystem," that can mean so much. So, I've narrowed it down to three factors, inspired by Joel Mokyr's 1992 book *The Lever of Riches: Technological Creativity and Economic Progress.*[243]

First is the social infrastructure. This goes back to my previous point on how people can only be entrepreneurial if they can take on risk. You can't expect anyone in sub-optimal conditions to think about billions of others when they struggle for their well-being. Thus, on an ecosystem level, we must improve basic resources around health care, nutrition, shelter, education, and more to get people focused in widespread change. We might need moonshots to fix those anyway.

Having an open and free society is also key to enabling progress because if you stimulate a higher level of collaboration and experimentation, the resulting radical creativity makes miracles. Specifically, this means applying proven

---

243 Mokyr, The Lever of Riches: Technological Creativity and Economic Progress.

frameworks and assumptions from one area to another uncertain one, almost like generalist thinking. If people can notice profound similarities across their specialized domains and join forces, all they need is capital and a beneficial sociopolitical environment to cultivate breakthroughs. Take down jargon barriers, recombine disparate fields, and build defensible moats as a result of it.[244]

Second, we need incentives. Not everyone has the intrinsic motivation to break the status quo or change their ways. The obvious go-to is more funding for labs and universities. As for nonobvious things, this may entail tax breaks on R&D: studies show reducing the price of R&D by 10 percent increases investment in innovation by 10 percent in the long run.[245] Or it could involve encouraging free trade and skilled migration: increasing the population of immigrant college graduates in the US by 1 percent increased patents per capita by up to 18 percent.[246] Other possibilities are protecting researchers' intellectual property, protecting employee benefits and rights, and encouraging healthy competition between companies.

We would also need self-sustaining projects. Huge problems like UN Sustainable Development Goals are not solved by one single entity, although that would be ideal. These crises need multiple "small" things. Thus, timescale plus easier innovation would yield a compounding impact. Take food scarcity. One single product cannot solve the entire issue, but the synergy of many organizations does wonders over time.

---

244 Sam Arbesman, interview by Benjamin Reinhardt.

245 Katia Savchuk, "The Five Best Policies to Promote Innovation — And One Policy to Avoid."

246 Ibid.

Now, there are nearly 820 million undernourished people worldwide, which is a major decline from past decades.[247] We still have much work to do across that mission-oriented sub-ecosystem.

Long-term thinking comes from having devoted participants—think religions and universities, both of which have been around for more than 1,000 years. In terms of business, Kongō Gumi was a Japanese construction company specializing in building temples. It operated for 1,400 years until its liquidation in 2006. The one thing we can learn from all of this is longevity, in this context, comes from adaptive, dynamic entities that are open to reinventing themselves.[248] It's about creating a cycle in which many people, even generations, are driven to do big things for a cause that gives back to them.

Doing all of this would normalize deeptech so it would be accessible for everyone. It would build a safety net and a herd-driven bias toward positive action spanning multiple industries and participants. The ripple effects of such efforts would be enormous.

Finally, we have the social attitude. In extremely simple terms, this is just the moonshot mindset in action throughout a larger collective. It's about being tolerant toward outsiders and accepting ambition. Almost like having a widespread growth mindset. For this, my conversation with Matt Clifford is most relevant. He is the founder of Entrepreneur First, a talent accelerator that spans almost every continent. The biggest cultural difference he sees between ecosystems that

247 Mercy Corps Editors, "The Facts: What You Need to Know about Global Hunger."

248 Sam Arbesman, interview by Benjamin Reinhardt.

churn out breakthroughs is the willingness to take risks. In fact, every single interview of mine yielded that same answer. It typically went: "Well, America is great because on average, being ambitious is encouraged. There is larger risk appetite." Interviewees described how England, Singapore, Japan, France, and other countries are far from that culturally, although some are on the come-up. Once the acts of setting inspiring-yet-intelligent goals and solving intractable problems are viewed as heroic, more deeptech startups will emerge.[249]

Taking in all of the nodes, contexts, and participants, the complex moonshot system clearly has many moving parts. Because people come and go in the dynamism of deeptech, we have to work together and learn from past mistakes to collectively create the future. Even with just one more attempt, all of society benefits. By social engineering a win-win ecosystem in which more people are radically creative, yet also have the concrete foundation to be, a golden age of *every* industry *will* come. We will be on our way to helping billions of people. Positive-sum interconnectedness is the panacea—the solution to everything.

We can finally reach Type I.

249 Steve Jurvetson, "Peter Thiel's Failed Futurism," Steve Jurvetson photostream (blog), Flickr, February 12, 2020.

6

CONCLUSION

WHEN MOONSHOTS
ARE MADE

"In ancient days, men looked at stars and
saw their heroes in constellations. In
modern times, we do much the same, but our
heroes are epic men of flesh and blood."

—WILLIAM SAFIRE,
(PRESIDENT RICHARD NIXON'S WRITER IN AN
ADDRESS TO THE NATION IN THE CHANCE THAT THE
APOLLO 11 ASTRONAUTS DIED ON THEIR JOURNEY.)

OPEN UP THE COSMOS
Doing big things is in our DNA. From Mesopotamia to charter
cities, and from the wheel to quantum supremacy, humanity's
pioneering spirit never ceases to exist. It's empowering—we
really can do anything.

In an American context, starting our country was thought
to be impossible. So were all of the advancements from the

multiple industrial revolutions. So was the Panama Canal. And, literally, so were the Apollo missions.

While the lunar term is a metaphor for doing the impossible, the motives behind such initiatives are not always directly for the betterment of all. The funny thing about the Apollo program was that it was not primarily an economic or scientific priority. It was instead a geopolitical one. See, JFK needed political backing considering Russia's advances in space. The institutions and capabilities we had were no match for such a feat—some people thought we needed to first get to low-earth orbit or at least establish a moon base before landing on it. But once again, humanity did its thing as it mobilized America into a mission-oriented goal that transcended many domains.

After Russian cosmonaut Yuri Gagarin's success as the first man to leave Earth's atmosphere, Kennedy remarked, "The first country that placed its national emblem on the moon was Russia, not America. The first passengers to return safely from a trip through space were named Strelka and Belka, not Rover or Fido." A 1960 Gallup poll found public sentiment in eight out of ten of the world's superpowers thought Russia was leading the world in deeptech and military prowess. The two dissenters in both studies were America and Greece. That didn't sit well with us. Say what you want about America, but nothing lights a fire under us like our competitiveness.[250]

We haven't gone back to the moon because the country became too bureaucratic, hyper-political, and indebted. And, it diverged from its commitment to advancing science and technology. There are rarely any political goals that run parallel with deeptech ones. It's only lip service. This is evidenced

250 Ryan Khurana, "The Political Machine Behind the Apollo Program."

by the tremendous difference in the military budget than R&D. Also, how ARPANET paved the way to the modern internet, but the War on Poverty did not sustain its efforts. Or, even Operation Warp Speed, which is making tangible progress in finding COVID-19 treatments, versus President Obama's Cancer Moonshot, which didn't do much to find a cure.[251] If there is a political incentive, we'll do it. If not, put it on the backburner.

I hate that mentality. That's why the private sector must stand up and build the things we need. Hence, my appreciation of audacious companies, not policies.

A CAMBRIAN EXPLOSION

There is a link, however, between the original moonshot, government, academia, and entrepreneurship.

One branch of this was the private-public sector collaboration between NASA and universities, namely MIT. The masterful management, the infectious social engineering, the precisely accurate manufacturing—NASA clearly had some commercial expertise. The culture of rigorous standards and perfect execution was rightfully so because with one mistake, human lives and $25.4 billion were at stake.[252] With JFK garnering political assets, NASA now needed the technology behind the missions.

This is where the other branch comes in. Space R&D had been going on for years prior with numerous relevant people, but this story starts with Nobel Prize winner William Shockley of Bell Labs, co-inventor of the transistor in

---

251 Healio Editors, "Despite Successful First Year, Cancer Moonshot's Mission 'Nowhere Near' Complete."

252 Alex Knapp, "Apollo 11's 50th Anniversary: The Facts and Figures Behind the $152 Billion Moon Landing."

1947. He then took his breakthrough from Murray Hill, New Jersey, to Mountain View, California, and founded Shockley Semiconductor in 1955. As a key figure within the computing industry, Shockley was widely known and attracted incredible talent. That said, his bitter management style turned off some of the scientists he hired. There were seven notable ones.[253]

Enter Arthur Rock, the legendary godfather of Silicon Valley. He is arguably the single most responsible person for Silicon Valley being what it is. As a top venture capitalist, and the first to use a limited-partner/general-partner structure, Rock serendipitously met some of those scientists who were disheartened by Shockley's leadership. Instead of getting them hired at another firm, he pushed them to start their own company.[254]

The group of seven added one more scientist as per a recommendation from Rock, and accordingly became "The Traitorous Eight." They came to an agreement with the existing company, Fairchild Camera and Instrument, to establish a semiconductor arm, which became known as the Fairchild Semiconductor. Thus, an industry giant was born. As for Rock, he went on to co-found, finance, or advise companies such as Teledyne, SDS, Intel, and Apple. No wonder he is considered the godfather of it all.[255]

The two branches now link together, as NASA and MIT needed advanced computing measures for their systems, specifically integrated circuits. Apollo became one of Fairchild Semiconductor's most important early customers, despite their technology being extremely novel and

---

253 Arthur Rock, Harvard Business School, accessed October 18, 2020.

254 Kevin Gao, "Letter #9: Arthur Rock (1987)," (blog).

255 Arthur Rock, Harvard Business School, accessed October 18, 2020.

untested on a massive scale. Together, the government-academia and venture-capital-startup synergy brought about a revolution, as the exigency and R&D yielded remarkably effective integrated circuits. NASA was not buying cheap ones. They were painfully quality-tested and quick to reject any batch of computer chips if even one was deemed faulty. They needed to be absolutely perfect. The weight of the Apollo missions rested on Fairchild's technology. It was a daunting task.[256]

But, as (almost) always, innovation pulled through. NASA and MIT asked, and Fairchild executed, making chips that skyrocketed computing power, reliability, and memory by nearly 1,000 times. To exemplify this, Gordon Moore, the theorist behind Moore's Law, coined the exponential concept explicitly based on NASA's operations. It was the perfect environment to study the rise of computational power in a real-world setting.[257]

The rest is history. Apollo 11 was a success. The impossible was impossible no more. But that's not all—the missions ignited other revolutions as well. One was in the digital sector, as the demand for ultra-precise computing and software trickled down into other industries. From the Apollo missions to now, the digital world has progressed so much I wouldn't even consider it deeptech anymore. It's heavily commoditized and holds trillions of dollars in value. Everything is becoming digitized. Next is the manufacturing realm, which was heavily boosted by Apollo's physical hardware requirements. At Fairchild, Texas Instruments, and many other Apollo suppliers, similar top-notch standards bled into manufacturing

256 Ryan Khurana, "The Political Machine Behind the Apollo Program."
257 Ibid.

plants. As a result, America boomed in hardtech: everything from automobiles to biotech.[258]

The imaginative and ambitious startup culture, the cutting-edge breakthroughs, and the resounding, captivating impact of it all—this what a moonshot is.

THAT'S ALL, FOLKS

We need more companies that emulate the positive sides of the Apollo program. That is, taking futuristic deeptech and making a radical solution out of it to solve a big problem in the world.

Every day, we are bombarded with bad news, saddening statistics, and media of wild horrors we have yet to solve. It's also the problems that go unnoticed yet affect millions. What most people do not realize, however, is a lot of it can be fixed with the plethora of deeptech we have or the uncommercialized research sitting in labs. To amplify all of this, an engaged ecosystem around a collective goal is needed. Crucial strategies are required, especially considering the ethical dilemmas and legal implications of science. We need to internalize philosophies such as maximizing human potential and mental models that cut through uncertainty. And the biggest part: it must be known that moonshots start in the mind. Altogether, the "moon" is the lofty goal we must reach and the "shot" is the execution.

But we can't just sit here and rhapsodize. Those who have the luxury and privilege of taking on entrepreneurial risk have to go out there and start these colossal projects. There's so much cool technology out there it's hard not to be optimistic about it. It's tough not to say, "Holy shit." We have

---

258 Ibid.

the beginnings of mind control, drugs that can cure almost every virus, smarter cities, and Mars exploration efforts. But we also have intractable problems. There are evident ones such as inadequate education and health care. But there are deeper, philosophical ones such as happiness and socioeconomic mobility, a lack of control over our existence, and a minimal understanding of the universe itself.

For younger people—we have to fix the notion science is boring. It's the exact opposite. It touches every aspect of life. My advice is to find things that excite you and double down on it—the money will follow. Balance it with soft sciences and humanities if that's where your passion lies. If you love video games, why not delve into augmented and virtual reality? If you love first-person shooters, why not learn about aerospace and defense technology? If you love sports, go into sports science, bionics, and tech-driven kinesiology. Fashion and makeup fanatics, there's so much talk around environmentally friendly clothes and cosmetology. Foodies and stoners, look into smarter agriculture and alternative medicines, such as how some psychedelics are effective for mental health. For my less active peers, if you love napping so much, explore hardware and software for sleep, especially the neuroscience side of it. There's so much out there in which you can make a considerable impact, get rich, and do what you love all at the same time. Especially for those of you whose blood boils at the sight of how messed up our world is—channel that energy into something like this.

My call to action: let's be heroes. Futurists. Myth-makers. Pioneers. Polymaths. We can bring forth a golden age that eradicates reckless deaths, bad economics, and universal suffering. Or, we can find humankind's fountain of youth in medicine, longevity science, and space travel.

I can give you all the metaphors of blissful utopias: renaissances and enlightenments, Shangri-las, El Dorados, and Type I civilizations. Unfortunately, nothing will matter if we just think about them and wait for the future to come. However, one thing we *can* do is build. Build bold things. The stuff of dreams and sci-fi, yet also those that add some good into the world.

Let's make some moonshots, through hardships and to the stars.

Per aspera ad astra.

ACKNOWLEDGMENTS

———

A huge thanks to the following people. You all made my dreams come true.

New Degree Press

Alan Zatkow	Brian Bies
Alexander Pyles	Eric Koester
Anne Belott	Tara Petronio

Beta Readers

Abhi Vyas	Jude Gomila
Alec Nevala-Lee	Kevin Gao
Anirudh Pai	Matt McGuire
Arnaud Schenk	Max Altman
Blas Moros	Ryan Bethencourt
Bradley Tusk	Seth Bannon
Dryden Brown	Will McCreadie
Jack Nasjaq	Will Robbins
Jordan Zoe Cobb	Zion Chan

Pre-sale Contributors

Aadit Kamat

Aahna Rathod

Aaron G. Baker

Alex Babayev

Alex LaBossiere-Barrera

Alexandra de Roos

Alisha Senapati

Ami Yoshimura

Amish Gupta

Amit Rana

Ana Sofia Almagro

Andrew Zhong

Anil & Bharti Rana

Anil Rana

Ankit Rana

Arnav Tolat

Arvind Farmah

Ashley Cha

Avaneen Pinninti

Benjamin Joffe

Bennett Liu

Brandon Goon

Camden De Sanctis

Carol Sheng

Chris Peckyno

Clayton Franklin

Cristian Rial

Cyrus Mowdawalla

Dakota Verrico

Daniel Luo

Deborah Hunke

Devan Patel

Devan Patel

Dhaval Patel

Dilan Bhavsar

Douglas Oliver

Edward Owen

Elaine Ku

Elena Andreacci

Elliot Lam

Emily Sun

Emma Scott

Enzo Ruta

Eric Chen

Eric Solorzano

Evan Bimaputra

Geetam Patchigolla

Gene Lin

Hannah Bondalo

Hannah Bondalo

Hemendra & Usha Rana

Hogene Park

Ikenna Onyekwere

Jack McClelland

Jackison Rana

Jagruti Ganshi

Janice Lee

Jatin Rana

Jay Rana

Jayesh Jariwala

Jigna Rana

Jignasa Rawalji

Jonathan Lee

Joseph Cobo

Kamlesh Rana

Kanaiya Hingu

Karan Kamara

Kathleen Steinberg

Kelvin Yu

Keyur Tankaria

Khushi Sabuwala

Krishna Talamati

Larry Drabeck

LaYow Yu

Lorenz Komischke

Maanav R. Arni

Malachi Sutton

Manju Ketan Shah

Manoj Rana

Mark Hammond

Matthew Clifford

Max Grev

Mia Bersalona

Mukti Rana-Singh

Munish Kohli

Muraldihar Manickam

Nagin Rana

Natasha Kanitkar

Neha Nandiwada

Neil Desai

Nicholas Lugo

Nick Frost

Nicole M. Bendik

Nidhi Shah

Nikhit Basineni

Nikita Nair

Noah Schneider

Paresh Patel

Paresh Rana

Pole Lee

Priya Rana

Priya Shah

Rahul Nandakumar

Rajesh Shah

Rajiv Singh

Rasmi Shah

Rayhan Murad

Rishub Nahar

Rithvik Reddy

Rohan Rana

Roshni Dhanasekar

Rucha Desai

Sandy Rana

Sanjay and Nita Rana

Sead Bajrovic

Sejal Rajagopalan

Shobha Dasari

Shreyas Ravindra

Soorya Das

Tharun Velishala

Thomas Edralin

Thomas Sully

Tina Chauhan

Tulip Sengupta

Twinkle Rana

Vaishali Rana

Varun Harvu

Vikas Rana

Viken Patel

Vishu Gupta

Will Robbins

Xerxes Kotval

Yan Leyzerovych

Zain Sohail

APPENDIX

INTRODUCTION

Redemption and Revolution

Davies, Alex. "Why 'Moon Shot' Has No Place in the 21st Century." *Wired*. July 16, 2019, https://www.wired.com/story/apollo-11-moonshot-21st-century/.

Friedman, Milton. "A Friedman Doctrine—the Social Responsibility of Business Is to Increase Its Profits." *The New York Times archive*. September 13, 1970. https://www.nytimes.com/1970/09/13/archives/a-friedman-doctrine-the-social-responsibility-of-business-is-to.html.

Frost, Jennifer. "Who Really Won the US-Soviet Space Race?" *Newsroom*. July 19, 2019. https://www.newsroom.co.nz/2019/07/19/688355/who-really-won-the-us-soviet-space-race.

History in HD. "JFK at Rice University, Houston, United States." Accessed October 22, 2020. https://unsplash.com/photos/rHIiT1Q6GIg.

Kay, Alan. "Early Meeting in 1971 of Parc, Palo Alto Research Center, Folks and the Xerox Planners." Speech presented at a meeting in Xerox PARC, Palo Alto, CA, 1971. https://quoteinvestigator.com/2012/09/27/invent-the-future/.

Kennedy, John F. "John F. Kennedy Moon Speech (1962)." AIRBOYD. Speech presented at Rice Stadium, Houston, TX, September 12, 1962. YouTube video, 17:47. https://youtu.be/TuW4oGKzVKc.

Madrigal, Alexis C. "Moondoggle: The Forgotten Opposition to the Apollo Program." *The Atlantic*, September 12, 2012. https://www.theatlantic.com/technology/archive/2012/09/moondoggle-the-forgotten-opposition-to-the-apollo-program/262254/.

National Aeronautics and Space Administration. "The Decision to Go to the Moon: President John F. Kennedy's May 25, 1961 Speech before a Joint Session of Congress." NASA History Office. Updated October 29, 2013. https://history.nasa.gov/moondec.html.

Ngo, Richard. "Thiel on Progress and Stagnation." LessWrong. July 20, 2020.
https://www.lesswrong.com/posts/Xqcorq5EyJBpZcCrN/thiel-on-progress-and-stagnation.

Osborne, Cary. "One of the First Los Angeles Dodgers Stars, Wally Moon Passes
Away." *Dodger Insider* (blog). *Medium,* February 10, 2018.
https://dodgers.mlblogs.com/one-of-the-first-los-angeles-dodgers-stars-wally-moon-
passes-away-ec0438b0ee2.

Teller, Astro. "We Choose to Go to the Moon." *X, The Moonshot
Factory* (blog). *Medium,* July 23, 2019.
https://blog.x.company/we-choose-to-go-to-the-moon-89596ef81e98.

MINDSET

The Blood of a Moonshot

ColdFusion. "The Wirecard Fraud - How One Man Fooled All of Germany." ColdFusion.
August 6, 2020. YouTube video, 17:05.
https://youtu.be/ivACzzW5wyA.

Csikszentmihalyi, Mihaly. *Creativity: Flow and the Psychology of Discovery and
Invention.* New York: Harper Perennial, 1997. Apple Books.

Elevate SIX. "Steve Jurvetson: Investing in Moonshot Companies." Elevate SIX. January
9th, 2020. YouTube video, 33:04.
https://youtu.be/eyt6sx-Koc8.

Fitzgerald, F. Scott. "The Crack-up: A Desolately Frank Document from One for Whom
the Salt of Life Has Lost Its Savor." *Esquire Classic.* February 1, 1936.
http://classic.esquire.com/the-crack-up/.

Frankish, Keith. "Our Greatest Invention Was the Invention of Invention
Itself." *Psyche.* June 24, 2020.
https://psyche.co/ideas/our-greatest-invention-was-the-invention-of-invention-itself.

Teller, Astro. "Tips for Unleashing Radical Creativity." *X, The Moonshot
Factory* (blog). *Medium,* February 12, 2020.
https://blog.x.company/tips-for-unleashing-radical-creativity-f4ba55602e17.

X Moonshot Factory. Gimbal_v2.0_X (PDF). Accessed August 29, 2020.

Y Combinator. "Elon Musk: How to Build the Future." Y Combinator. September 15,
2016. YouTube video, 19:32.
https://www.youtube.com/watch?v=tnBQmEqBCY0&t=6s.

Iconoclast Inventors

"Beal Aerospace Fires Largest Liquid Rocket Engine in 30 Years." *Spaceref.* March 4, 2000.
http://www.spaceref.com/news/viewpr.html?pid=1047.

Chaddha, Navin. "Biology as Technology Will Reinvent Trillion-Dollar
Industries." *TechCrunch.* September 17, 2019.
https://techcrunch.com/2019/09/17/biology-as-technology-will-reinvent-trillion-
dollar-industries/.

Greely, Henry T. "The Two Months in 1980 That Shaped the Future of Biotech." *Stat on Apple News.* October 17, 2020. https://apple.news/AaH7JTe3WRVy4n6xHOkL96A.

Kofinas, Demetri. "Neural Interfaces and the Future of Human-Computer Interaction | Thomas Reardon." Interview of Thomas Reardon by Demetri Kofinas. *Hidden Forces,* March 11, 2019. Audio, 1:04:57. https://podcasts.apple.com/us/podcast/hidden-forces/id1205359334?i=1000431518446.

Kuhn, Thomas S. *The Structure of Scientific Revolutions: 50th Anniversary Edition.* Chicago: University of Chicago Press, 2012. Apple Books.

Mendell, Lorne M. "The Size Principle: A Rule Describing the Recruitment of Motoneurons." *Journal of Neurophysiology,* vol. 93 (Summer 2005): https://journals.physiology.org/doi/pdf/10.1152/classicessays.00025.2005.

Rigden, John S. "Einstein's Revolutionary Paper." *Physicsworld.* April 1, 2005. https://physicsworld.com/a/einsteins-revolutionary-paper/.

Rodriguez, Salvador. "Facebook Agrees to Acquire Brain-Computing Startup CTRL-Labs." *CNBC.* September 23, 2019. https://www.cnbc.com/2019/09/23/facebook-announces-acquisition-of-brain-computing-startup-ctrl-labs.html.

Smith, Jonathan. "Humble Beginnings: The Origin Story of Modern Biotechnology." *Labiotech.eu.* July 2, 2019. https://www.labiotech.eu/in-depth/history-biotechnology-genentech/.

"Thomas Kuhn: The Structure of Scientific Revolutions." *Farnam Street* (blog). *fs. blog,* Accessed August 16, 2020. https://fs.blog/2011/09/how-scientific-advancement-happens/.

Weinstein, Eric. "The Mind Financing the Future." Interview of Josh Wolfe by Eric Weinstein. *The Portal*, May 3rd, 2020. Audio, 1:42:37. https://podcasts.apple.com/us/podcast/the-portal/id1469999563?i=1000473458522.

Wolfe, Josh. "Episode 1: The Magic of Your Mind; CTRL-labs' Neural Interface Unlocks Human Potential."

Lux Capital. August 1, 2019. YouTube Video, 6:28. https://youtu.be/Iuhrs8UbDRQ.

A Golden Age for Dreams

Breakthrough Initiatives. "Starshot." https://breakthroughinitiatives.org/initiative/3.

"Early Meeting in 1971 of Parc, Palo Alto Research Center, Folks and the Xerox Planners." Speech presented at a meeting in Xerox PARC, Palo Alto, CA, 1971. https://quoteinvestigator.com/2012/09/27/invent-the-future/.

Hello Tomorrow. "In the Future We'll Wear Spider Silks | Dan Widmaier | HT Summit 2017." Hello Tomorrow. November 20, 2017. YouTube video, 15:59. https://youtu.be/4KsRv6OBHBw.

Kurzweil, Ray. "The Law of Accelerating Returns." *Kurzweil Accelerating Intelligence* (blog). *Essays*, March 7, 2001. https://www.kurzweilai.net/the-law-of-accelerating-returns.

Maples Jr., Mike. "How to Build a Breakthrough." Mike Maples Jr. on Medium (blog). *Medium,* April 27, 2020. https://medium.com/@m2jr/how-to-build-a-breakthrough-3071b6415b06.

Norberg, Johan. "Our Golden Age." *The Spectator.* August 20, 2016. https://www.spectator.co.uk/article/our-golden-age.

Patrick, G.T.W. "The New Optimism." *The Popular Science Monthly* 82, (1913): 492–494. https://archive.org/details/popularsciencemo82newy/page/492/mode/2up.

Solomon, Lisa Kay. "How Leaders Dream Boldly to Bring New Futures to Life." *SingularityHub.* February 23, 2017. https://singularityhub.com/2017/02/23/how-leaders-dream-boldly-to-bring-new-futures-to-life/.

Theil, Peter. *Zero to One: Notes on Startups, or How to Build the Future.* New York, NY: Currency, 2014. http://morfene.com/021.pdf.

Audacity Can Be Easy

Charter Cities Institute. "An Introduction to Charter Cities." Accessed August 30, 2020 https://www.chartercitiesinstitute.org/intro.

Girihlet. "What We Do." Accessed August 30, 2020. https://www.girihlet.com/what-we-do.html.

Leng, Sidney. "China's Tech Hub Shenzhen Set to Hit 2019 Growth Target after Surprising Fourth Quarter Rebound Amid Trade War." *South China Morning Post.* January 9, 2020. https://www.scmp.com/economy/china-economy/article/3045400/chinas-tech-hub-shenzhen-set-hit-2019-growth-target-after.

Taleb, Nassim Nicholas. *Antifragile: Things That Gain from Disorder.* New York, NY: Random House, 2012. Apple Books.

Teller, Astro. "The Head of 'X' Explains How to Make Audacity the Path of Least Resistance." *Wired,* April 15, 2016. https://www.wired.com/2016/04/the-head-of-x-explains-how-to-make-audacity-the-path-of-least-resistance/.

Trading Economics. "Hong Kong GDP." Accessed August 30, 2020. https://tradingeconomics.com/hong-kong/gdp.

Trading Economics. "Singapore GDP." Accessed August 30, 2020. https://tradingeconomics.com/singapore/gdp.

X Moonshot Factory. Gimbal_v2.0_X (PDF. Accessed August 30, 2020.

Two Is Better Than One

Brondmo, Hans Peter. "Robotics Is a Team Sport." *X, The Moonshot Factory* (blog). *Medium,* June 5, 2019. https://blog.x.company/robotics-is-a-team-sport-4bf2edd65871.

Brown, Matthew. "Fact Check: Bill Gates Has Given over $50 Billion to Charitable Causes over Career." *USA Today.* June 11, 2020. https://www.usatoday.com/story/news/factcheck/2020/06/11/fact-check-bill-gates-has-given-over-50-billion-charitable-causes/3169864001/.

"Ernest Shackleton's Crew of the Endurance Imperial Trans Antarctica Expedition 1914 -17." Cool Antarctica. Accessed August 26, 2020. https://www.coolantarctica.com/Antarctica%20fact%20file/History/antarctic_whos_who_endurance.php.

History.com Editors. "Cubism History." *History.* Updated August 21, 2018. https://www.history.com/topics/art-history/history-of-cubism.

Jurvetson, Steve. "Moore Evermore in Computer History — Happy 50th Birthday to the Law!" *Steve Jurvetson photostream* (blog). *Flickr.* February 11, 2015. https://www.flickr.com/photos/jurvetson/16318918399/.

Teller, Astro. "Tips for Unleashing Radical Creativity." *X, The Moonshot Factory* (blog). *Medium,* February 12, 2020. https://blog.x.company/tips-for-unleashing-radical-creativity-f4ba55602e17.

"The Stormtrooper Problem: Why Thought Diversity Makes Us Better." *Farnam Street* (blog). *fs.blog.* Accessed August 26, 2020. https://fs.blog/2019/03/stormtrooper-problem/.

PHILOSOPHY

More Celestial Power to You

Thaller, Michelle. "We Are Dead Stars: We Are Born of Supernovas – Our Spectacular and Totally Ordinary Origin Story." September 14, 2017. In *Aeon.* Produced by Flora Lichtman and Katherine Wells. Video, 3:56. https://aeon.co/videos/we-are-born-of-supernovas-our-spectacular-and-totally-ordinary-origin-story.

The Moral Imperative for Moonshots

Frankl, Viktor E. *Man's Search for Meaning.* Boston: Beacon Press, 1946. https://edisciplinas.usp.br/pluginfile.php/3403095/mod_resource/content/1/56ViktorFrankl_Mans%20Search.pdf.

Frankl, Viktor E. *Man's Search for Meaning.* Boston: Beacon Press, 1946. https://edisciplinas.usp.br/pluginfile.php/3403095/mod_resource/content/1/56ViktorFrankl_Mans%20Search.pdf.

O'Shaughnessy, James. "Josh Wolfe – The Tech Imperative." April 23, 2019. In *Invest Like the Best.* Produced by The Investor's Field Guide. Podcast, MP3 audio, 1:17:15. https://podcasts.apple.com/us/podcast/invest-like-the-best/id1154105909?i=1000436137469.

Viktor Frankl Institute of Logotherapy. "Logotherapy." Accessed September 9, 2020. http://www.logotherapyinstitute.org/About_Logotherapy.html.

Master the Mind to Master the World

Aristotle. *Nichomacean Ethics, Book I, Chapter 4.* Translated by W.D. Ross. http://classics.mit.edu/Aristotle/nicomachaen.1.i.html.

Encyclopedia Britannica Online. Academic ed s.v. "Ancient Stoicism." Accessed September 11, 2020, https://www.britannica.com/topic/Stoicism/Ancient-Stoicism.

Epictetus. *The Enchiridion.* Translated by Elizabeth Carter. http://classics.mit.edu/Epictetus/epicench.html.

Ferriss, Tim. "Stoicism 101: A Practical Guide for Entrepreneurs." *The Tim Ferriss Show* (blog). *tim.blog,* April 13, 2009. https://tim.blog/2009/04/13/stoicism-101-a-practical-guide-for-entrepreneurs/.

"Introduction to Effective Altruism." *Effective Altruism.* June 22, 2016. https://www.effectivealtruism.org/articles/introduction-to-effective-altruism/.

Moore, Catherine. "What Is Eudaimonia? Aristotle and Eudaimonic Well-Being." *Positive Psychology.* January 9, 2020. https://positivepsychology.com/eudaimonia/.

Nobel Media. "Norman Borlaug — Facts." *NobelPrize.org.* Accessed October 11, 2020. https://www.nobelprize.org/prizes/peace/1970/borlaug/facts/.

Srinivasan, Balaji S. "The Purpose of Technology." *Balaji S. Srinivasan* (blog). *Ghost,* July 19, 2020. https://balajis.com/the-purpose-of-technology/.

Thompson, Derek. "The Greatest Good." *The Atlantic.* June 15, 2015. https://www.theatlantic.com/business/archive/2015/06/what-is-the-greatest-good/395768/.

The Phoenix Within

Azhar, Azeem. "Inside Alphabet's X: Nurturing Radical Creativity." Interview of Astro Teller by Azeem Azhar. Exponential View with Azeem Azhar, Harvard Business Review, April 8, 2020. Audio, 38:06. https://podcasts.apple.com/us/podcast/exponential-view-with-azeem-azhar/id1172218725?i=1000470882943.

Cardon, Melissa S., Joakim Wincent, Jagdip Singh, and Mateja Drnovsek. "The Nature and Experience of Entrepreneurial Passion." *Academy of Management Review* 34, no. 3 (2009): 511–532. https://pdfs.semanticscholar.org/ea51/cd276d206e409096c5b71fb52c4854fce173.pdf.

Dandelion. "Environmental Impact." Accessed September 21, 2020. https://dandelionenergy.com/environmental-impact.

Felten, Obi. "How to Kill Good Things to Make Room for Truly Great Ones." *X, The Moonshot Factory* (blog). *Medium,* March 8, 2016. https://blog.x.company/how-to-kill-good-things-to-make-room-for-truly-great-ones-867fb6ef026.

Huckman, Robert S., Karim R. Lakhani, and Kyle R. Myers. March 2018. *X: The Foghorn Decision: Case 618-060.* Distributed by Harvard Business School.
https://www.hbs.edu/faculty/Pages/item.aspx?num=54400.

MacMillan, Douglas. "Astro Teller, 'Captain of Moonshots' at Alphabet's X, Is on a Roll." *Wall Street Journal,* October 23, 2018.
https://www.wsj.com/articles/astro-teller-captain-of-moonshots-at-alphabets-x-is-on-a-roll-1540310778.

Teller, Astro. "Tips for Unleashing Radical Creativity." *X, The Moonshot Factory* (blog). *Medium,* February 12, 2020.
https://blog.x.company/tips-for-unleashing-radical-creativity-f4ba55602e17.

"The Role of Error in Innovation." *Farnam Street* (blog). *fs.blog,* Accessed September 21, 2020.
https://fs.blog/2012/11/the-role-of-error-in-innovation/.

Moonshot Mental Models

Altman, Sam. "How to Be Successful." *Sam Altman* (blog). *blog.samaltman,* January 24, 2019.
https://blog.samaltman.com/how-to-be-successful.

Aristotle. *Metaphysics.* Quoted in: "First Principles: The Building Blocks of True Knowledge." *Farnam Street* (blog). *fs.blog,* Accessed October 11, 2020.
https://fs.blog/2018/04/first-principles/.

Arthur, W. Brian. "Increasing Returns." *W. Brian Arthur* (blog). *Sante Fe Institute,* 2018.
http://tuvalu.santafe.edu/~wbarthur/increasingreturns.htm.

"Compounding Knowledge." *Farnam Street* (blog). *fs.blog,* Accessed October 11, 2020.
https://fs.blog/2019/02/compounding-knowledge/.

"First Principles: The Building Blocks of True Knowledge." *Farnam Street* (blog). *fs. blog,* Accessed October 11, 2020.
https://fs.blog/2018/04/first-principles/.

Karlgaard, Rich. "Wriston's Law Still Holds." *Forbes.* July 17, 2009.
https://www.forbes.com/sites/digitalrules/2009/07/17/wristons-law-still-holds/#7e2d946c312e.

Kelly, Kevin. "68 Bits of Unsolicited Advice." *The Technium* (blog). *kk,* April 28, 2020.
https://kk.org/thetechnium/68-bits-of-unsolicited-advice/.

Kevin Rose. "Foundation 20 // Elon Musk." Kevin Rose. September 7, 2012. YouTube video, 26:42.
https://youtu.be/L-s_3b5fRd8.

Lee, Stan. *Amazing Fantasy (1962) #15.* Penciled by Steve Ditko. New York City, NY: Marvel Comics, 1962.

"Mental Models: The Best Way to Make Intelligent Decisions." *Farnam Street* (blog). *fs. blog,* Accessed October 11, 2020.
https://fs.blog/mental-models/.

Moros, Blas. "Advantageous Divergence." *The Latticework.* Accessed October 11, 2020. https://ltcwrk.com/worldly-wisdom/advantageous-divergence/.

Moros, Blas. "Second-Order Thinking." *The Latticework.* Accessed October 11, 2020. https://ltcwrk.com/worldly-wisdom/second-order-thinking/.

Moros, Blas. "Systems Thinking." *The Latticework.* Accessed October 11, 2020. https://ltcwrk.com/worldly-wisdom/systems-thinking/.

Ravikant, Naval. "Make Luck Your Destiny." *Naval* (blog). *nav.al,* March 7, 2019. https://nav.al/luck-destiny.

Scott, Ridley, dir. *The Martian.* 2015; Los Angeles, CA: 20th Century Fox, 2015. Cinema.

Tierney, John and Roy F. Baumeister. "For the New Year, Say No to Negativity." *The Wall Street Journal.* December 27, 2019. https://www.wsj.com/articles/for-the-new-year-say-no-to-negativity-11577464413.

Weinberg, Gabriel. "Mental Models I Find Repeatedly Useful." *Gabriel Weinberg* (blog). *Medium,* July 5, 2016. https://medium.com/@yegg/mental-models-i-find-repeatedly-useful-936f1cc405d.

Wolfe, Josh. 2017. *Josh Wolfe_volume2.* Kinetic Energy Ventures: Compilations. https://drive.google.com/file/d/1ldRf9JomaRsA_S-g8V6TQPkBOujO5UJ7/view?usp=sharing.

STRATEGY

The Future That Never Was

Churchill, Winston. "Fifty Years Hence." *Popular Mechanics,* March 1932. http://rolandanderson.se/Winston_Churchill/Fifty_Years_Hence.php.

X Moonshot Factory. Gimbal_v2.0_X (PDF). Accessed October 14, 2020.

We're Gonna Need a Bigger Boat

Davé, Neil. "Introducing Tidal." *X, The Moonshot Factory* (blog). *Medium*, March 2, 2020. https://blog.x.company/introducing-tidal-1914257962c3.

Hebert, Peter and Josh Wolfe. "From Wasteland to Fund-Maker." *Lux Capital* (blog), *Medium,* February 3, 2016. https://medium.com/lux-capital/from-wasteland-to-fund-maker-4c7aab08c45f.

History.com Editors. "Manhattan Project." *History.* Updated January 17, 2020. https://www.history.com/topics/world-war-ii/the-manhattan-project.

Jain, Naveen. "Exponential Thinking with Naveen Jain." Interview by Jim Kwik. *Kiwk Brain with Jim Kwik,* June 6, 2018. Audio, 19:48. https://jimkwik.com/kwik-brain-059-exponential-thinking-with-naveen-jain/.

Lux Capital. "Kurion Overview." Lux Capital. January 31, 2016. YouTube video, 3:19. https://youtu.be/Hsa8R_BRK0o.

Portincaso, Massimo, Arnaud de la Tour, and Philippe Soussan. "The Dawn of the Deep Tech Ecosystem." *Boston Consulting Group.* March 14, 2019. https://www.bcg.com/publications/2019/dawn-deep-tech-ecosystem.

X Moonshot Factory. "Tidal." Accessed October 14, 2020. https://x.company/projects/tidal/.

Moonshots Are for Everyone

Azhar, Azeem. "Inside Alphabet's X: Nurturing Radical Creativity." Interview of Astro Teller by Azeem Azhar. Exponential View with Azeem Azhar, Harvard Business Review, April 8, 2020. Audio, 38:06. https://podcasts.apple.com/us/podcast/exponential-view-with-azeem-azhar/id1172218725?i=1000470882943.

Capital Camp. "Josh Wolfe - The Future, Now." Capital Camp. February 17, 2020. YouTube video, 51:45. https://youtu.be/mZm81yus4mQ.

Henderson, J. Maureen. "Meet the Teen Who Got Paid $100,000 to Drop Out of School." *Forbes.* June 20, 2011. https://www.forbes.com/sites/jmaureenhenderson/2011/06/20/meet-the-teen-who-got-paid-100-000-to-drop-out-of-school/#26c742666dc2.

Jain, Naveen. "Why Non-experts Are Better at Disruptive Innovation." *Naveen Jain Blog* (blog). *naveenjain.com*, May 12, 2015. http://www.naveenjain.com/why-non-experts-are-better-at-disruptive-innovation/.

"Mike Maples: Living in the Future [The Knowledge Project Ep. #77]." *Farnam Street* (blog). *fs.blog,* Accessed October 14, 2020. https://fs.blog/knowledge-project/mike-maples/.

Rabang, Imelda. "The Airbnb Startup Story: An Odd Tale of Airbeds, Cereal and Ramen." *Bold Business.* May 31, 2019. https://www.boldbusiness.com/society/airbnb-startup-story/.

Tom Bilyeu. "Naveen Jain on Why Curiosity Will Save the World | Impact Theory." Impact Theory. July 11, 2017. YouTube video, 50:25. https://youtu.be/Y_itHlMwrFA.

Money Talks

Denning, Steve. "The Science of Storytelling." *Forbes.* March 9, 2012. https://www.forbes.com/sites/stevedenning/2012/03/09/the-science-of-storytelling/#60a37c302d8a.

Leiva, Ludmila. "Keep Track of the Theranos Scandal with This Detailed Timeline." *Refinery29.* January 29, 2019. https://www.refinery29.com/en-us/2019/01/222855/theranos-scandal-timeline-what-happened-elizabeth-holmes-documentary.

Ott, Tim. "Inside Elizabeth Holmes and the Downfall of Theranos." *Biography.* Updated June 4, 2020. https://www.biography.com/news/elizabeth-holmes-theranos-scam.

Roff, Connie. "Everything You Need to Know about the Theranos Scandal." *Pan Macmillan*. March 3, 2019.
https://www.panmacmillan.com/blogs/literary/theranos-elizabeth-holmes-john-carreyrou.

Psihoyos, Louie, dir. *The Game Changers*. 2019; USA: Netflix, 2019. Netflix.

Zack Schildhorn. "Lux Capital's Zack Schildhorn on the Power of Narrative for Entrepreneurs." Lux Capital. January 22, 2020. YouTube video, 8:36.
https://youtu.be/-llii7frwco.

The Bleeding-Edge

Bush, Vannevar. "Science: The Endless Frontier." *United States Government Printing Office*. July 1945.
https://www.nsf.gov/od/lpa/nsf50/vbush1945.htm.

Evans, Kelly. "The Great Stagnation, Low-Hanging Fruit and America's 'Sputnik Moment'." *The Wall Street Journal*. January 31, 2011.
https://blogs.wsj.com/economics/2011/01/31/the-great-stagnation-low-hanging-fruit-and-americas-sputnik-moment/.

Gupta, Arvind. "The $100 Trillion Opportunity." *IndieBio* (blog). *Medium,* January 16, 2019.
https://medium.com/indiebio-sf/the-100-trillion-opportunity-3d827f18c56a.

Khosla, Vinod. "Reinventing Societal Infrastructure with Technology." *Vinod Khosla* (blog). *Medium,* April 22, 2018.
https://medium.com/@vkhosla/reinventing-societal-infrastructure-with-technology-f71e0d4f2355.

Taylor, Bill. "What Breaking the 4-Minute Mile Taught Us about the Limits of Conventional Thinking." *Harvard Business Review*. March 9, 2020.
https://hbr.org/2018/03/what-breaking-the-4-minute-mile-taught-us-about-the-limits-of-conventional-thinking.

Sci-Fi Meets Sci-Fact

Cameron, James, dir. *The Terminator*. 1984; Los Angeles, CA: Orion Pictures, 1984. Blu-ray Disc, 1080p HD.

Capital Camp. "Josh Wolfe - The Future, Now." Capital Camp. February 17, 2020. YouTube video, 51:45.
https://youtu.be/mZm81yus4mQ.

CB Insights Editors. "Game Changing Startups 2019." *CB Insights*. 2019.
https://www.cbinsights.com/reports/CB-Insights_Game-Changers-2019.pdf.

Goodreads. "William Gibson Quotes." Accessed October 14, 2020.
https://www.goodreads.com/quotes/681-the-future-is-already-here-it-s-just-not-evenly.

Klus, Helen. "Imagining the Future: Why Society Needs Science Fiction." *The Star Garden* (blog), *thestargarden.co.uk,* Updated January 1, 2020.
http://www.thestargarden.co.uk/Why-society-needs-science-fiction.html.

Kubrick, Stanley, dir. *2001: A Space Odyssey*. 1968; Beverly Hills, CA: Metro-Goldwyn-Mayer, 1968. Blur-ray Disc, 1080p HD.

Lee, Stan. *The X-Men: Return of the Blob (1964) #7.* Penciled by Jack Kirby. New York City: Marvel Comics, 1964.

Lucas, George, dir. *Star Wars: Episode I – The Phantom Menace.* 1999; Los Angeles, CA: 20th Century Fox, 1999. Blu-ray Disc, 1080p HD.

Lucas, George, dir. *Star Wars: Episode IV – A New Hope.* 1977; Los Angeles, CA: 20th Century Fox, 1977. Blu-ray Disc, 1080p HD.

Lucas, George, dir. *Star Wars: Episode V – The Empire Strikes Back.* 1980; Los Angeles, CA: 20th Century Fox, 1980. Blu-ray Disc, 1080p HD.

"Mental Models: The Best Way to Make Intelligent Decisions (109 Models Explained)." *Farnam Street* (blog). *fs.blog,* Accessed October 14, 2020. https://fs.blog/mental-models/.

Roddenberry, Gene, exec. prod. *Star Trek: The Original Series.* Aired September 8, 1966 – June 3, 1969, on NBC.

Roddenberry, Gene and Rick Berman, exec. prod. *Star Trek: The Next Generation.* Aired September 28, 1987 – May 23, 1994, first-run syndication.

Rosen, Rebecca J. "Why Today's Inventors Need to Read More Science Fiction." *The Atlantic.* September 20, 2013. https://www.theatlantic.com/technology/archive/2013/09/why-todays-inventors-need-to-read-more-science-fiction/279793/.

Verhoeven, Paul, dir. *Total Recall.* 1990; Culver City, CA: TriStar Pictures, 2990. Blu-ray Disc, 1080p HD.

Zemeckis, Robert, dir. *Back to the Future Part II.* 1989; Universal City, CA: Amblin Entertainment Universal Pictures, 1989. Blu-ray Disc, 1080p HD.

ECOSYSTEM

Diamond Age

Kanisetti, Anirudh. "What Makes a Golden Age?" *Indian National Interest* (blog). *Medium,* October 11, 2017. https://nationalinterest.in/what-makes-a-golden-age-7daef5ed7d01.

Martin, Roger. "The Virus Shows That Making Our Companies Efficient Also Made Our Country Weak." *The Washington Post.* March 27, 2020. https://www.washingtonpost.com/outlook/2020/03/27/economic-efficiency-resilience-coronavirus/.

Pai, Anirudh. "Golden Age & Polis." *Athwart.* June 10, 2020. https://www.athwart.org/golden-age-and-polis/.

Reinhardt, Benjamin. "Type I and Type II Progress." Benjamin Reinhardt (blog), *benjaminreinhardt.com,* January 3, 2020. https://benjaminreinhardt.com/type-I-and-type-II-progress/.

Pegasus, Not Unicorn

Chad Rigetti. "Chad Rigetti at Startup School SV 2016." Y Combinator. September 29, 2016. YouTube video, 24:01.
https://youtu.be/GzMvG8UO6Eg.

Comstock, Douglas A. and Daniel Lockney. "NASA's Legacy of Technology Transfer and Prospects for Future Benefits." *NASA.* 2007.
https://www.nasa.gov/pdf/330843main_aiaa_space_2007.pdf.

Currier, Jamies. "The Hidden Patterns of Great Startup Ideas." *NFX.* Accessed October 16, 2020.
https://www.nfx.com/post/hidden-patterns-great-startup-ideas/.

Halioua, Celine. "Applying Tech Frameworks to Biotech: Key Differences." *Celine Halioua* (blog). *celinehh.com,* Accessed October 16, 2020.
https://www.celinehh.com/tech-vs-biotech.

NASA. NASA Spinoff 2020 (PDF). Accessed October 16, 2020.
https://spinoff.nasa.gov/Spinoff2020/pdf/Spinoff2020.pdf.

NASA/Airspace Systems. "NASA Technological Readiness Level Chart." Accessed October 22, 2020.
https://commons.wikimedia.org/wiki/File:NASA_TRL_Meter.png.

X Moonshot Factory. "Wing." Accessed October 16, 2020.
https://x.company/projects/wing/.

Money Makes the World Go Round

Clark, Kate. "US VC Investment in Female Founders Hits All-Time High." *TechCrunch.* December 9, 2019.
https://techcrunch.com/2019/12/09/us-vc-investment-in-female-founders-hits-all-time-high/.

Different Funds. "DeepTech Investing Report." *Different.* March 2020.

Foust, Jeff. "Advanced Space Propulsion Startup Shuts Down." *Spacenews.* February 5, 2016.
https://spacenews.com/advanced-space-propulsion-startup-shuts-down/.

Loria, Kevin. "This Man's Potentially Huge Medical Breakthrough Can't Get Funding, So He's Trying Something Desperate." *Business Insider.* December 15, 2020.
https://www.businessinsider.com/todd-rider-draco-crowdfunding-broad-spectrum-antiviral-2015-12.

O'Shaughnessy, James. "Josh Wolfe – The Tech Imperative." April 23, 2019. In *Invest Like the Best.* Produced by The Investor's Field Guide. Podcast, MP3 audio, 1:17:15.
https://podcasts.apple.com/us/podcast/invest-like-the-best/
id1154105909?i=1000436137469.

The Engine. "2019 Tough Tech Landscape." *The Engine.* 2019.

The Old Guard in New Times

Anduril. Careers. Accessed October 17, 2020.
https://www.anduril.com/careers.

Bush, Vannevar. "Science: The Endless Frontier." *United States Government Printing Office*. July 1945.
https://www.nsf.gov/od/lpa/nsf50/vbush1945.htm.

Congressional Research Service. "Federal Research and Development (R&D) Funding: FY2020." *Congressional Research Service*. Updated March 18, 2020.

Different Funds. "DeepTech Investing Report." *Different*. March 2020.

Douthat, Ross. *The Decadent Society: How We Became the Victims of Our Own Success*. New York, NY: Avid Reader Press / Simon & Schuster, 2020.

Farmwave. Homepage. Accessed October 17, 2020.
https://www.farmwave.io/.

Hourihan, Matt and David Parkes. "Federal R&D Budget Trends: A Short Summary." *American Association for the Advancement of Science*. January 2019.
https://www.aaas.org/sites/default/files/2019-01/AAAS%20R%26D%20Primer%202019.pdf.

Jurvetson, Steve. "Peter Thiel's Failed Futurism" *Steve Jurvetson photostream* (blog). *Flickr*. February 12, 2020.
https://www.flickr.com/photos/jurvetson/49527474492.

Meyer, Michael. "The Rise and Fall of Vannevar Bush." *Science History*. July 21, 2018.
https://www.sciencehistory.org/distillations/the-rise-and-fall-of-vannevar-bush.

National Constitution Center. Preamble. Accessed October 26, 2020.
https://constitutioncenter.org/interactive-constitution/preamble.

O'Hanlon, Michael E. "Is US Defense Spending Too High, Too Low, or Just Right?" *Brookings*. October 15, 2019.
https://www.brookings.edu/policy2020/votervital/is-us-defense-spending-too-high-too-low-or-just-right/.

Pai, Anirudh. "Between Atoms and Bits." *Dreams of Electric Sheep* (blog). *Substack,* September 27, 2020.
https://electricsheep.substack.com/p/5c753525-6f5b-4bef-88c7-20fe552aa66f.

An Engine for Learning

Eric Weinstein. "Eric Weinstein (Solo), Ep. #018 of The Portal - Slipping the DISC: State of The Portal/Chapter 2020." Eric Weinstein. February 13, 2020. YouTube video, 1:03:35.
https://youtu.be/QxnkGymKuuI.

Fitzgerald, Jay. "Does Science Advance One Funeral at a Time?" *National Bureau of Economic Research*. Accessed October 17, 2020.

Jain, Naveen. "Naveen Jain on Rethinking Education: Can We Use Neuroscience to Create Better Learners." *Forbes*. June 16, 2011.
https://www.forbes.com/sites/naveenjain/2011/06/16/naveen-jain-on-rethinking-education-can-we-use-neuroscience-to-create-better-learners/#3e662c873c00.

Martin, Karen. "Publishing Failure in Science." *GoldBio*. Accessed October 17, 2020.
https://www.goldbio.com/articles/article/Publishing-Failure-in-Science.

Okamura, Keisuke. "Interdisciplinarity Revisited: Evidence for Research Impact and Dynamism." *Palgrave Commun* 5, no. 141 (Fall 2019). https://doi.org/10.1057/s41599-019-0352-4.

Capitalizing on Complexity

Anthony, Scott D. "How Singapore Became an Entrepreneurial Hub." *Harvard Business Review.* February 25, 2015. https://hbr.org/2015/02/how-singapore-became-an-entrepreneurial-hub.

Arbesman, Sam. "Compounding Ideas with Sam Arbesman [Idea Machines #16]." Interview by Benjamin Reinhardt. *Idea Machines,* Benjamin Reinhardt, May 24, 2019. Audio, 53:12. https://ideamachinespodcast.com/sam-arbesman.

Different Funds. "DeepTech Investing Report." *Different.* March 2020.

Jurvetson, Steve. "Peter Thiel's Failed Futurism" *Steve Jurvetson photostream* (blog). *Flickr.* February 12, 2020. https://www.flickr.com/photos/jurvetson/49527474492.

Mercy Corps Editors. "The Facts: What You Need to Know about Global Hunger." *Mercy Corps* (blog). *mercycorps.org*, May 8, 2020. https://www.mercycorps.org/blog/quick-facts-global-hunger.

Mokyr, Joel. *The Lever of Riches: Technological Creativity and Economic Progress.* Oxford: Oxford University Press, 1992.

Savchuk, Katia. "The Five Best Policies to Promote Innovation — And One Policy to Avoid." *Insights by Stanford Business.* October 7, 2019. https://www.gsb.stanford.edu/insights/five-best-policies-promote-innovation-one-policy-avoid.

WatchMojo. "What If Humanity Was a Type II Civilization? | Unveiled." Unveiled. July 22, 2020. YouTube video, 8:01. https://youtu.be/y6Aj_bnZ3Gs.

CONCLUSION

When Moonshots Are Made

Gao, Kevin. "Letter #9: Arthur Rock (1987)." *A Letter A Day* (blog). *Substack,* July 15, 2020. https://aletteraday.substack.com/p/letter-9-arthur-rock-1987.

Harvard Business School. Arthur Rock. Accessed October 18, 2020.

Healio Editors. "Despite Successful First Year, Cancer Moonshot's Mission 'Nowhere Near' Complete." *Healio.* May 25, 2017. https://www.healio.com/news/hematology-oncology/20170515/despite-successful-first-year-cancer-moonshots-mission-nowhere-near-complete.

Khurana, Ryan. "The Political Machine Behind the Apollo Program." *Palladium.* July 15, 2020. https://palladiummag.com/2020/07/15/the-political-machine-behind-the-apollo-program/.

Knapp, Alex. "Apollo 11's 50th Anniversary: The Facts and Figures Behind the $152 Billion Moon Landing." *Forbes.* July 20, 2019. https://www.forbes.com/sites/alexknapp/2019/07/20/apollo-11-facts-figures-business/#5396a42e3377.

Printed in Great Britain
by Amazon